Bibliographic information published by the German National Library:

The German National Library lists this publication in the National Bibliography; detailed bibliographic data are available on the Internet at http://dnb.dnb.de .

Imprint:

Copyright © 2018 GRIN Verlag
Print and binding: Books on Demand GmbH, Norderstedt Germany
ISBN: 9783668818972

This book at GRIN:

https://www.grin.com/document/445009

Matthias Himmelmann

Galois Groups and Fundamental Groups on Riemann Surfaces

GRIN Verlag

Freie Universität Berlin

Department of Mathematics and Computer Science

Research Project for obtaining a Bachelor's Degree in Mathematics

Galois Groups and Fundamental Groups on Riemann Surfaces

Matthias Himmelmann

Winter Semester 2017/18

March 5, 2018

Introduction

For a sufficiently nice topological space X we will see that the fundamental group $\pi_1(X, x_0)$ at a point $x_0 \in X$ is isomorphic to the group of deck transformations of the universal covering $p : E \to X$, i.e. homeomorphisms σ of E such that the following diagram commutes:

$$\begin{array}{ccc} E & \xrightarrow{\sigma} & E \\ & {\scriptstyle p}\searrow \quad \swarrow{\scriptstyle p} & \\ & X & \end{array}$$

Similarly, for a field X and a *separable closure* E of X, the Galois group of X is nothing but the set of automorphisms σ of E that fix X. In other words, those are all automorphisms on E, such that the following diagram commutes:

$$\begin{array}{ccc} E & \xrightarrow{\sigma} & E \\ & {\scriptstyle \iota}\nwarrow \quad \nearrow{\scriptstyle \iota} & \\ & X & \end{array}$$

Here, $\iota : X \to E$ denotes the field inclusion map.

This should already give us the idea that there is a correlation between the fundamental group and the Galois group, even though we might not be able to fully grasp how this correlation might look like.

To do this, our initial action will be to recall some basic definitions and properties from category theory, algebra and topology that will be necessary for introducing other notions later on. The main goal of this thesis will be to prove certain properties of the category of finite covering spaces and the category of some special k-algebras over a field k and to eventually compare them. For that and for seeing similiarities, it will be useful to also prove properties for the objects of these categories.

The last chapter gives an explicit example for when fundamental groups and Galois groups occur simultaneously. For a *Riemann surface* X, i.e. a 1-dimensional complex manifold, it is possible to construct an abstract correlation between finite field extensions of its field of meromorphic functions $\mathcal{M}(X)$ and *branched coverings*, which are nothing but special coverings. This correlation needs the language of category theory. The final theorem of this thesis will give a more explicit correlation, actually an isomorphism, between the Galois group of a particular field extension and the *profinite completion* of the fundamental group of the punctured space, which is a kind of limit of the fundamental group modulo its normal subgroups of finite index. This is meant to put the fundamental group and the Galois group into a context.

Another aim of this thesis will be to compare the *separable closure* k^{sep} of a field k and the *universal cover* $\tilde{X}$ of a topological space X. One will notice that it will be possible to put the subgroups of the groups of automorphisms on $\tilde{X}$ that fix X (the deck transformations) into a one-to-one correspondence with the *subordinate* covering spaces, which sounds quite similar to the well-known *Fundamental Theorem of Galois Theory*. However, this theorem only holds for finite field extensions, we will expand it to infinite Galois extensions. In doing so, we will see a correspondence between $Aut(\tilde{X}/X)$ and $Gal(k^{sep}/k)$ and thus between the universal cover and the separable closure.

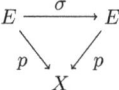

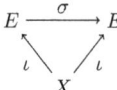

Table of Contents

Chapter 1

Algebraic Foundations

Before beginning with outlining the concepts that this thesis strives to cover, let us recall some basic concepts that will be necessary throughout the thesis. The aim of this chapter is to summarize known concepts and therefore will not be shaped by a common thread, but rather bringing back to mind some algebraic notions.

1.1 Category Theory

Definition 1.1.1 A *category* $\mathcal{C}$ consists of a class of objects $ob(\mathcal{C})$ and a class $Hom(\mathcal{C})$ of morphisms between those objects. Given two objects A and B, we write $Hom_{\mathcal{C}}(A, B)$ for the set of morphisms $A \to B$. We require:

1. for $\phi \in Hom_{\mathcal{C}}(A, B)$ and $\psi \in Hom_{\mathcal{C}}(B, D)$ there is $\psi \circ \phi \in Hom_{\mathcal{C}}(A, D)$ and we call it *composition*. For this, we require associativity.

2. $\forall A \in ob(\mathcal{C})\ \exists\ id_A \in Hom_{\mathcal{C}}(A, A)$, the *identity morphism*, that fulfils $\phi \circ id_A = \phi = id_B \circ \phi$ for any $\phi \in Hom_{\mathcal{C}}(A, B)$.

In the following, we will only deal with *small categories*, meaning that the objects of this category form a set.

Definition 1.1.2 For a category $\mathcal{C}$, the *opposite category* $\mathcal{C}^{op}$ is defined by $ob(\mathcal{C}^{op}) = ob(\mathcal{C})$ and reversing the morphisms, so we have $Hom_{\mathcal{C}}(A, B) = Hom_{\mathcal{C}^{op}}(B, A)$ for any $A, B \in ob(\mathcal{C})$.

Definition 1.1.3 In a category $\mathcal{C}$, an object Z is called *final*, if for each object B there is a unique morphism $B \to Z$.
Conversely, an object is called *initial*, if for each object B there is a unique morphism $A \to B$.

Example: After having seen some definitions, let us introduce a first example for a better understanding of the topic. We view the small category $\mathcal{C}$ with objects A, B, D and morphisms that are depicted in the following diagram. Here, the identity morphisms are assumed but not included.

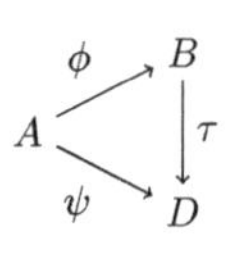

In the picture we can see that A is the initial object of $\mathcal{C}$ and D is the final object of $\mathcal{C}$.

Definition 1.1.4 A *(covariant) functor* F between two categories $\mathcal{C}$ and $\mathcal{D}$ is a mapping that associates to each object X an object $F(X)$ and to each morphism $g : X \to Y$ in $Hom(\mathcal{C})$ a morphism $F(g) : F(X) \to F(Y)$ in $Hom(\mathcal{D})$, such that F preserves identity morphisms and composition of morphisms.
Conversely, a *contravariant functor* from $\mathcal{C}$ to $\mathcal{D}$ is a functor from $\mathcal{C}$ to $\mathcal{D}^{op}$.

Definition 1.1.5 In a category $\mathcal{C}$, a morphism $f : A \to B$ is called a *monomorphism*, if for any $g, h : D \to A$, $f \circ g = f \circ h$ implies that $g = h$.

Apart from that, a morphism $f : A \to B$ is called an *epimorphism*, if for any $j, k : B \to D$, $j \circ f = j \circ f$ implies that $j = k$.

Finally, we call $f : A \to B$ an *isomorphism*, if there exists a morphism $g : B \to A$ such that $f \circ g = id_B$ and $g \circ f = id_A$.

Example: In the category of *sets*, the morphisms are functions. Furthermore, we can easily see that the monomorphisms are injective functions, the epimorphisms are surjective functions and isomorphisms are bijective functions.

Definition 1.1.6 Let $\mathcal{C}_1$ and $\mathcal{C}_2$ be categories and F, G functors $\mathcal{C}_1 \to \mathcal{C}_2$. A *natural transformation* Ψ is a collection of morphisms $\psi_A : F(A) \to G(A) \in Hom(\mathcal{C}_2)$ for each object A in $\mathcal{C}_1$ such that for any morphism $\phi : A \to B \in Hom(\mathcal{C}_1)$ the following diagram commutes:

$$\begin{array}{ccc} F(A) & \xrightarrow{\psi_A} & G(A) \\ {\scriptstyle F(\phi)}\downarrow & & \downarrow{\scriptstyle G(\phi)} \\ F(B) & \xrightarrow{\psi_B} & G(B) \end{array}$$

The natural transformation Ψ is an *isomorphism*, if each ψ_A is an isomorphism and in this case we say that F and G are isomorphic.

Definition 1.1.7 Two categories $\mathcal{C}_1$ and $\mathcal{C}_2$ are called *equivalent*, if there exist two covariant functors $F : \mathcal{C}_1 \to \mathcal{C}_2$ and $G : \mathcal{C}_2 \to \mathcal{C}_1$ and two natural transformations that are isomorphisms $\Psi : F \circ G \to id_{\mathcal{C}_2}$ and $\Phi : G \circ F \to id_{\mathcal{C}_1}$.
Otherwise, we call $\mathcal{C}_1$ and $\mathcal{C}_2$ *anti-equivalent*, if $\mathcal{C}_1$ is equivalent to $\mathcal{C}_2^{op}$.

Example: Consider the category *Grp* of groups with group homomorphisms as morphisms. Let $(G, \star)$ be a group. Then $(G^{op}, \star^{op})$ is its opposite group, where $\star^{op}$ is defined as $a \star^{op} b = b \star a$ and G and G^{op} have the same underlying set. If we define for any group homomorphism $f : G \to H$ the opposite group homomorpism to be $f^{op} = f : G^{op} \to H^{op}$, then we get

$$f^{op}(a \star^{op} b) = f^{op}(b \star a) = f(b \star a) = f(b) \star f(a) = f^{op}(b) \star f^{op}(a) = f^{op}(a) \star^{op} f^{op}(b)$$

This induces a covariant functor ${}^{op} : Grp \to Grp$ that maps a group to its opposite group. This is naturally isomorphic to id_{Grp}. To prove that, by definition 1.1.6 we need to find a collection of group isomorphisms $\psi_G : G^{op} \to G$ such that for any group homomorphism $\phi : G \to H$ it holds that $\phi \circ \psi_G = \psi_H \circ \phi^{op}$. We will see that the inversion proves this to be true. By choosing $\psi_G(g) = g^{-1}$, since the inverse element is unique in a group and the inversion is obviously a group homomorphism. Its inverse map is ψ_G^{op}, therefore it is an isomorphism. To prove that it is a natural transformation, we show

$$\psi \circ \phi_G(g) = \psi(g^{-1}) = \psi(g)^{-1} = \phi_H \circ \psi(g) = \phi_H \circ \psi^{op}(g)$$

Consequently the natural transformation between ^{op} and id_{Grp} is an isomorphism. As a result, by definition 1.1.7 *Grp* and *Grp* are equivalent categories, which is not really a surprising result.

Definition 1.1.8 Let $F : \mathcal{C}_1 \to \mathcal{C}_2$ be a functor. It is *faithful*, if for any $A, B \in ob(C_1)$ it holds that the map $F_{AB} : Hom_{\mathcal{C}_1}(A, B) \to Hom_{\mathcal{C}_2}(F(A), F(B))$ is injective. It is called *fully faithful*, if the maps F_{AB} are bijective for any two given objects.
Opposite to this, the functor $F : \mathcal{C}_1 \to \mathcal{C}_2$ is called *essentially surjective*, if every object of $\mathcal{C}_2$ is isomorphic to some object of the form $F(A)$.

Lemma 1.1.9 *Two categories $\mathcal{C}_1$ and $\mathcal{C}_2$ are anti-equivalent if there exists a contravariant functor from $\mathcal{C}_1$ to $\mathcal{C}_2$ that is fully faithful and essentially surjective.*

PROOF Let us first notice that this would be a functor $F : \mathcal{C}_1 \to \mathcal{C}_2^{op}$, so let us fix, using the essential surjectivity of F for any $V \in ob(\mathcal{C}_2)$ an isomorphism $i_V : F(A) \tilde{\to} V$ for some $A \in ob(\mathcal{C}_1)$. Then define a functor $G : \mathcal{C}_2^{op} \to \mathcal{C}_1$ by $G(V) = A$ and $G(\phi : V \to W) = F_{AB}^{-1}(i_W^{-1} \circ \phi \circ i_V)$, where $B = G(W)$ and F_{AB} is the bijection that is induced by the fully faithfulness of F (see 1.1.8). The maps $i_V : F(G(V)) \tilde{\to} V$ then induce the isomorphism $\Phi : F \circ G \tilde{\to} id_{\mathcal{C}_2^{op}}$ from definition 1.1.6.

For constructing the second isomorphism $\Psi : G \circ F \tilde{\to} id_{\mathcal{C}_1}$ that is by definition 1.1.7 necessary for the anti-equivalence of categories, we need to construct $\Psi_A : G(F(A)) \to A$ for each $A \in ob(\mathcal{C}_1)$. Since F is fully faithful by assumption, it suffices to construct a map $F(\Psi_A) : F(G(F(A))) \to F(A)$ and we may take the unique preimage of $id_{F(A)}$ under Φ as $F(\Psi_A)$. Similarly, we can construct a map $\Psi_A^{-1} : A \to G(F(A))$ that is inverse to Ψ_A, thus yielding us the desired isomorphism.
□

Definition 1.1.10 (Pullback and Fiber Product, [3])
Let $\mathcal{C}$ be a category and let $f : A \to C$ and $g : B \to C$ be morphisms. A *pullback* of f and g consists of morphisms p_1 and p_2 such that $g \circ p_2 = f \circ p_1$ and such that this has the following *universal property*: for all morphisms $x_1 : Y \to A$ and $x_2 : Y \to B$ such that $f \circ x_1 = g \circ x_2$, there exists a unique morphism $u : Y \to P$ such that $x_1 = p_1 \circ u$ and $x_2 = p_2 \circ u$. This ensures that a pullback is unique up to a canonical isomorphism. Equivalently to this definition, we can say that the following diagram commutes:

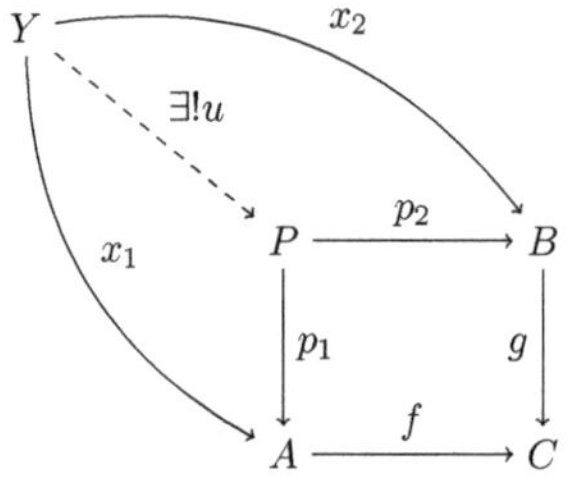

We call P the *fiber product* of A and B over C and denote it by $P = A \times_C B$.

Lemma 1.1.11 *Let A, B, C be sets and let $\xi : A \to C$ and $\mu : B \to C$ be two maps. Then the fibre product of A and B over C is*

$$A \times_C B = \{(a, b) \in A \times B : \xi(a) = \mu(b)\} \tag{1.1}$$

PROOF First, we need to find maps p_1 and p_2, but we can choose them to be the restrictions of the two projections $\pi_A : A \times B \to A$ and $\pi_B : A \times B \to B$ to $A \times_C B$. If we pick any $(a, b) \in A \times_C B$, then

$$\xi \circ \pi_A(a, b) = \xi(a) \overset{(1.1)}{=} \mu(b) = \mu \circ \pi_B(a, b)$$

Since we picked an arbitrary $(a, b) \in A \times_C B$, this implies that $\xi \circ \pi_A = \mu \circ \pi_B$.

Now, let Y be another object with morphisms $x_1 : Y \to A$ and $x_2 : Y \to B$ such that $\xi \circ x_1 = \mu \circ x_2$. This implies that u is uniquely determined by $u(y) = (x_1(y), x_2(y))$. This is well-defined by the fact that $\xi(x_1(y)) = \mu(x_2(y))$. □

Definition 1.1.12 Let $(X_i)_{i \in I}$ be a collection of objects in $\mathcal{C}$. The *coproduct* of this collection is an object $\bigsqcup_{i \in I} X_i$ together with the morphisms $q_j : X_j \to \bigsqcup_{i \in I} X_i$ for each $j \in I$, such that for any object Y of $\mathcal{C}$ and any collection of morphisms $f_j \in Hom_{\mathcal{C}}(X_j, Y)$, $j \in I$, there is a unique morphism $f : \bigsqcup_{i \in I} X_i \to Y$ such that $f_j = f \circ q_j$ for all $j \in I$.
In other words, for a coproduct, we expect the following diagram to commute for any chosen $j, j' \in I$:

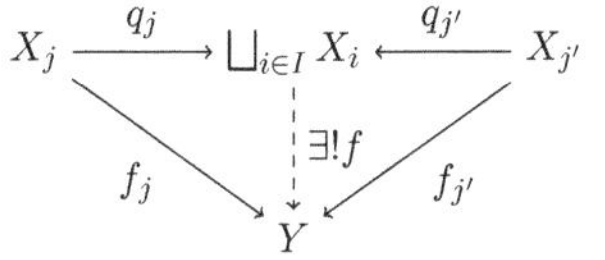

Definition 1.1.13 Let X be an object in $\mathcal{C}$ and let G be a subgroup of the group of isomorphisms in $Hom_{\mathcal{C}}(X, X)$ (*automorphisms*). The *categorial quotient* of X by G is an object X/G in $\mathcal{C}$ with a morphism $p : X \to X/G$ in $\mathcal{C}$ such that for all $\sigma \in G$ we have $p = p \circ \sigma$. Additionally, we require that for any morphism $f \in Hom_{\mathcal{C}}(X, Y)$ with some object Y that satisfies $f = f \circ \sigma$ for all $\sigma \in G$, there exists a unique $g : X/G \to Y$ such that $f = g \circ p$. The following commutative diagram sums this up:

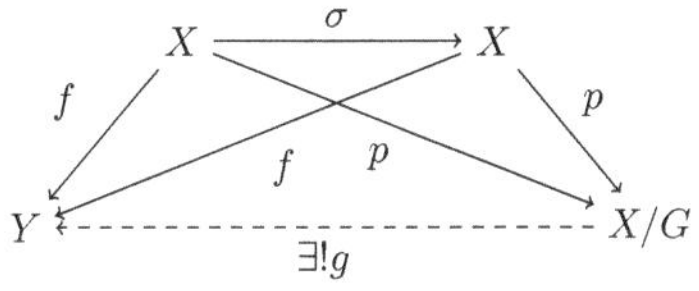

Lemma 1.1.14 *If we are in a category where the objects are sets, the coproduct is the usual disjoint union together with the inclusion maps ι_j with $\iota_j(x) = (x, j)$:*

$$\bigsqcup_{i \in I} X_i = \bigcup_{i \in I} \{(x, i) : x \in X_i\}$$

The categorial quotient of X by a group of automorphisms G is then

$$X/G = \{\{g(x) : g \in G\} : x \in X\}$$

PROOF For the first part of the lemma, let us choose for any X_j a map $f_j : X_j \to Y$ for some object Y. We define f on every disjoint set X_j to be f_j as follows:

$$f(x, i) = f_i(x)$$

This way we get for each $j \in I$ that $f_j = f \circ \iota_j$.

On the other hand, for the second part of the proof we need to check that X/G with the topological quotient map $p : X \to X/G$. Here, the property $p = p \circ \sigma$ for all $\sigma \in G$ is clear from the definition. Now, let $f : X \to Y$ be a function that satisfies $f = f \circ \sigma$ for any $\sigma \in G$. We first notice that by this property f is constant on the equivalence classes of X/G. Accordingly, we can define g by restricting f to X/G. □

Remark: This lemma and lemma 1.1.11 will from now on be our standard tools for proving properties for categories whose objects are sets.

Definition 1.1.15 A morphism $u : X \to Y$ is called a *direct summand*, if there is a morphism $q_2 : Z \to Y$ such that Y together with $q_1 = u$ and q_2 is the coproduct $X \sqcup Z$ of X and Z.

1.2 Profinite Groups

Definition 1.2.1 Let G be a topological space whose underlying set is equipped with a structure of a group with operation $\star$. We call G a *topological group*, if the group operation $(a, b) \mapsto a \star b$ and inversion $a \mapsto a^{-1}$ are continuous maps.

Definition 1.2.2 Let $(I, \prec)$ be a non-empty set I together with a binary relation $\prec$. It is called a *directed set*, if it fulfills the following properties:

1. $\prec$ is reflexive: For any $i \in I$ it holds that $i \prec i$.
2. $\prec$ is antisymmetric: For any $i, j \in I$ with $i \prec j$ and $j \prec i$ it holds that $i = j$.
3. $\prec$ is transitive: For all $i, j, k \in I$ with $i \prec j$ and $j \prec k$ it holds that $i \prec k$.
4. $\prec$ is directed: For all $i, j \in I$ there exists a $k \in I$ that satisfies $i \prec k$ and $j \prec k$.

Definition 1.2.3 Let $(I, \prec)$ be a directed set and $(S_i)_{i\in I}$ a collection of sets indexed by I. Now suppose for each $j \prec i$ there exists a map $\mu_{ij} : S_i \to S_j$ satisfying the following conditions:

$$\begin{aligned} \mu_{ji} \circ \mu_{kj} &= \mu_{ki} && \text{for all } i \prec j \prec k \\ \mu_{ii} &= id_{S_i} && \text{for all } i \in I \end{aligned}$$

These three entities together are called a *projective system*. The *projective limit* $\varprojlim S_i$ of this projective system is then defined to be

$$\varprojlim S_i = \left\{ (x_i)_{i\in I} \in \prod_{i\in I} S_i \ : \ \mu_{ij}(x_i) = x_j \text{ for all } i, j \in I \text{ with } j \prec i \right\}$$

Proposition 1.2.4 (Profinite Groups, [5])
Let $((I, \prec), (G_i)_{i\in I}, (\mu_{ij})_{j\prec i})$ be a projective system, where all G_i are finite groups and μ_{ij} are group homomorphisms. Then $G = \varprojlim G_i$ is a group.
If each G_i is equipped with the discrete topology and $\prod_i G_i$ with the product topology then G is a closed topological subgroup of $\prod_i G_i$.

Proof We use the fact that $\pi = \prod_{i\in I} G_i$ is a group and show that the projective limit G from definition 1.2.3 is a subgroup of that group. Since G_i are all groups, it holds that the identity element $1_{G_i} \in G_i$ and, since the μ_{ij} are group homomorphisms it holds that $\mu_{ij}(1_{G_i}) = 1_{G_j}$ for all $j \prec i$, which implies that G is non-empty.
Let $(a_i)_{i\in I}$, $(b_i)_{i\in I}$ be two elements of G. Then

$$(a_i)_{i\in I} \star_\pi (b_i)_{i\in I} = (a_i \star_{G_i} b_i)_{i\in I}$$

and since all the μ_{ij} are group homomorphisms, we have for any pair of $j \prec i$:

$$\mu_{ij}(a_i \star_{G_i} b_i) = \mu_{ij}(a_i) \star_{G_j} \mu_{ij}(b_i) = a_j \star_{G_j} b_j$$

Hence, $(a_i)_{i\in I} \star_\pi (b_i)_{i\in I} \in G$ which implies that G is closed under $\star_\pi$.
Now, let $(a_i)_{i\in I} \in G$ be an arbitrary element. Using that μ_{ij} are group homomorphisms, a similar argument shows that $(a_i)_{i\in I}^{-1} \in G$ which implies that G is closed under inversion.
Ultimately, this shows that G is a subgroup of π, yielding that G is a group itself.

Recall that π is equipped with the product topology. The open basis elements are of the form

$$\prod_{j\in S} U_j \times \prod_{i\in I\setminus S} G_i \qquad \text{for a finite } S \subset I$$

Here, $U_j \subset G_j$ are open sets and as each G_j is equipped with the discrete topology, the U_j are an arbitrary subset of G_j. By definition 1.2.1 a group is a topological group, if both, the group operation and inversion, are continuous. Since G is equipped with the subspace topology, it suffices to show that the group operation and inversion are continuous as viewed on π, since on G they are nothing but a restriction from π.

For this matter, let us choose an arbitrary $(x_j)_{j\in S} \in \prod_{j\in S} G_j$ for some finite set $S \subset I$. We denote $U_{(x_j)_{j\in S},S} = \prod_{j\in S}\{x_j\} \times \prod_{i\in I\setminus S} G_i$. Since this forms an open basis for the profinite group, it suffices to check continuity for $U_{(x_j)_{j\in S},S}$. We prove that for any $(a_i)_{i\in I} \times (b_i)_{i\in I} \in \pi^2$ and any open basis neighorbood V of $(a_i)_{i\in I} \star_\pi (b_i)_{i\in I}$ we find a neighborhood of $(a_i)_{i\in I} \times (b_i)_{i\in I}$ that $\star_\pi$ maps entirely into V. So let $V = U_{((a_i)\star_{G_i}(b_i))_{i\in S},S}$ for some finite $S \subset I$. Then, obviously, $\star_\pi(U_{(a_i)_{i\in S},S} \times U_{(b_i)_{i\in S},S}) \subset V$, since the operation $\star_\pi$ is the operation $\star_{G_i}$ on each coordinate $i \in I$, which implies that $\star_\pi$ is continuous at $(a_i)_{i\in I} \times (b_i)_{i\in I}$ and since this construction can equivalently be done at any point $(a_i)_{i\in I} \times (b_i)_{i\in I} \in \pi^2$, we know that $\star_\pi$ is continuous altogether. A similar construction with $(x_i^{-1})_{i\in I} \in \pi$ indicates that the inversion $(\ )^{-1}$ is also continuous. Ultimately, the restrictions of $\star_\pi$ and $(\ )^{-1}$ to G are also continuous which implies that G is a topological group.

In addition, G is closed in $\pi = \prod_i G_i$, because when choosing a point $(x_i)_{i\in I} \in \pi$ where $\mu_{ij}(x_i) \neq x_j$ for some $j \prec i$ let us pick the open basis set that contains x_i and x_j, namely $U_{x_j} \cap U_{x_i}$. It is open in the product topology, since intersections of open sets are open. Moreover, it is disjoint from G by construction. We can conclude that G is closed. □

Definition 1.2.5 Following the above poposition we call $\varprojlim G_i$ a *profinite group*.

Furthermore, for an arbitrary group G we can view the I as the set of all normal subgroups of finite index of G. Equipped with the order relation $N \prec M \Leftrightarrow N \supset M$ we observe that $(I, \prec)$ is a directed set. Then $G_N = G/N$ with $N \in I$ are groups and we get the natural group homomorphisms $\mu_{NM} : G_N \to G_M$ with $\mu_{NM}(x) = Mx$ for $M \prec N$. This is a projective system by 1.2.3. Then the inverse limit $\hat{G} = \varprojlim G/N$ is called the *profinite completion* of G.

Remark: There is a natural homomorphism $\mu : G \to \hat{G}$ for the profinite completion $\hat{G}$ of G that is characterized in the following way: Given any profinite group H and any group homomorphism $f : G \to H$, there exists a unique and continuous group homomorphism $g : \hat{G} \to H$ such that $f = g \circ \mu$.

Example $[\hat{\mathbb{Z}}]$: Let us view the projective system consisting of $\mathbb{Z}/n\mathbb{Z}$ for $n \geq 1$. In this case, the partial order is given by the divisibility relation $m|n \Leftrightarrow m\mathbb{Z} \supset n\mathbb{Z}$ which is a directed set, since every two integers have a least common multiple. The group homomorphisms are $\mu_{nm} : \mathbb{Z}/n\mathbb{Z} \to \mathbb{Z}/m\mathbb{Z}$ that sends $x \mapsto x \bmod m$. It is well-defined, if $m|n$. The profinite completion $\hat{\mathbb{Z}} = \varprojlim \mathbb{Z}/n\mathbb{Z}$ contains a copy of $\mathbb{Z}$ by taking $(0, 1, 1, 1, \dots)$ to be 1 and the rest to be generated by the representative of 1 through the natural addition of the rings $\mathbb{Z}/n\mathbb{Z}$. Each $m(0, 1, 1, 1, \dots)$ is unique for $m \in \mathbb{Z}$, because there are infinitely many $\mathbb{Z}/n\mathbb{Z}$.

Additionally, if we take a non-empty open basis set $U = (\prod_{m\in S}\{x_m\} \times \prod_{n\in\mathbb{N}\setminus S} \mathbb{Z}/n\mathbb{Z}) \cap \hat{\mathbb{Z}}$ for a finite set $S \subset \mathbb{N}$ and $x_m \in \mathbb{Z}/m\mathbb{Z}$, we shall prove that there is an element of the form $k(0, 1, 1, \dots)$ in U for $k \in \mathbb{Z}$. By shrinking U to a new open set U', if necessary, we can assume that the newly constructed S' is of the form $\{m : m|N\}$ for some $N \in \mathbb{N}$, e.g. $N = lcm(S)$. This is well-defined, since S is finite. The construction is possible, since we can just choose $S' = S \cup \{N\}$ and $U' = U \cap (\{x_N\} \times \prod_{n\in\mathbb{N}\setminus\{N\}} \mathbb{Z}/n\mathbb{Z})$. The open set U is non-empty by assumption, so we just pick $x_N \in \mathbb{Z}/N\mathbb{Z}$ that is compatible with the group homomorphisms, which is possible by the definition of $\hat{\mathbb{Z}}$. Then we found a natural number x_N that has the requested property $x_N(0, 1, 1, \dots) \in U' \subset U$, since the group homomorphisms μ_{Nm} map $y \mapsto y \ mod \ m$, so we get by the compatibility and the fact $x_N \in U'$ that $\mu_{Nm}(x_N) = x_N \ mod \ m = x_m$ for any $m \in S'$. This yields that $\mathbb{Z}$ is a dense subset of $\hat{\mathbb{Z}}$.

Definition 1.2.6 Let $(G, \star_G)$ be a group and X a set. A *group action* ϕ of G on X is a function

$$\phi : G \times X \to X, \ (g, x) \mapsto g \cdot_\phi x$$

where $\cdot_\phi$ satisfies $e \cdot_\phi x = x$ and $g \cdot_\phi (h \cdot_\phi x) = (g \star_G h) \cdot_\phi x$. We then call X a *G-set*. Furthermore, the action is called *free*, if for all non-trivial $g \in G$ and all $x \in X$ it holds that $g \cdot_\phi x \neq x$ and it is called *trivial*, if for all $g \in G$ and $x \in X$ it holds that $g \cdot_\phi x = x$.
If E is a G-set, we denote $E^G = \{e \in E : g \cdot_\phi e = e \text{ for all } g \in G\}$ the set of *G-invariants* of E. In general, it is neither free nor trivial.

Example: Let G be a group and let H be a subgroup of G. Then multiplication from the left is an action of G on the set of cosets G/H, since $g(aH) = (ga)H$ for all $g, a \in G$. By implication, $e_G(aH) = (e_G a)H = aH$ for all $a \in G$.

1.3 Finite Field Extensions

Definition 1.3.1 Consider two fields $E \supset F$. This is called a *field extension* of F and we denote it by E/F. This way, E becomes a F-vector space and we call the dimension of this vector space the *degree of the extension* and denote it by $[E : F]$. In this manner, a *finite extension* E/F is a field extension of finite degree.
An *F-automorphism* is a field automorphism on E that preserves F. It is easy to verify that the F-automorphisms form a group together with composition. We denote this group by $Aut(E/F)$. In addition, the field extension E/F is called *algebraic*, if every element $\alpha \in E$ is the root of some non-zero polynomial $f \in F[X]$. Thus, we call α algebraic over F.

Lemma 1.3.2 *Let $L/E/F$ be a tower of field extensions. L/F is of finite degree if and only if L/E and E/F are of finite degree. If this is the case, the following equation holds:*

$$[L : F] = [L : E] \cdot [E : F]$$

PROOF If L is of finite degree over F, it most certainly is of finite degree over E. Moreover, E, being a subspace of a finite dimension vector space, is also of finite dimension.
Now, let $(f_j)_{j=1,\dots,n}$ be the basis of E as an F-vector space and $(e_i)_{i=1,\dots,m}$ the basis of L as an E-vector space. It is sufficient to show that $(e_i f_j)_{i=1,\dots,m;\ j=1,\dots n}$ is a basis of L as a F-vector space because n and m are finite and so is $n \cdot m$.
Firstly, we prove that $(e_i f_j)$ generates L. For some $l \in L$ we have that

$$l = \sum_{i=1}^{m} \alpha_i e_i \qquad \text{for some } \alpha_i \in E.$$

Just as E is an F-vector space, so

$$\alpha_i = \sum_{j=1}^{n} \beta_{i,j} f_j \qquad \text{for some } \beta_{i,j} \in F.$$

The logical conclusion is that

$$l = \sum_{i=1}^{m} \sum_{j=1}^{n} \beta_{i,j} e_i f_j$$

and thus $(e_i f_j)$ generates L.
Secondly, $(e_i f_j)$ is linear independent. This is due to the fact that a linear relation $\sum \beta_{i,j} e_i f_j = 0$, $\beta_{i,j} \in F$ can be rearranged to $\sum_i (\sum_j \beta_{i,j} f_j) e_i = 0$. Since (e_i) forms a basis, it is linear independent and therefore $\sum_j \beta_{i,j} f_j = 0$ for each i. By the same token, (f_j) form a basis and are consequently linear independent. This implies that $\beta_{i,j} = 0$ and that is why $(e_i f_j)$ is linear independent. □

Definition 1.3.3 Let F be a field. A non-constant polynomial is called irreducible over F, if its coefficients belong to F and it cannot be factored into the product of two non-constant polynomials with coefficients in F.

Definition 1.3.4 Let E/F be an algebraic field extension. For some element $\alpha \in E$ the *minimal polynomial* m_α of α is a monic generator of the ideal $\{h \in F[X] : h(\alpha) = 0\}$.

Lemma 1.3.5 *For an algebraic field extension E/F and any $\alpha \in E$ the minimal polynomial m_α is irreducible.*

PROOF Suppose m_α was not irreducible. From definition 1.3.4 we know that m_α is non-constant, since the only constant polynomial that has α as a zero is the constant zero polynomial. As α is algebraic over F by definition 1.3.1 there is a non-constant polynomial that has α as a root which is why the ideal, call it I_α, from definition 1.3.4 is non-zero. This implies that the generator of said ideal is non-constant (note: $c \notin I_\alpha$ for any $c \in F^\star$).
Applying that m_α is not irreducible, it can be factored into two non-constant polynomials, denote them by f and g, with coefficients in F. Now, α is a root of m_α, so it has to be the root of either f or g. Without loss of generality, let us assume that α is a root of f. Hence, m_α cannot generate f, since $deg(f) < deg(m_\alpha)$, but we know $f(\alpha) = 0$, so $f \in I_\alpha$. This contradicts to the fact that m_α generates I_α and implies that m_α is irreducible. □

Definition 1.3.6 An algebraic field extension E/F is called *normal*, if every irreducible polynomial $f \in F[X]$ either has no root in E or splits into linear factors in E.
The extension E/F is called *separable*, if for any $\alpha \in E$ the minimal polynomial $m_\alpha \in F[X]$ has distinct roots in an algebraic closure of F. Such a polynomial is called *separable*.

Definition 1.3.7 If an algebraic field extension E/F is both, separable and normal, we call it a *Galois extension*. We call $Aut(E/F)$ the *Galois group* of the extension E/F and we denote it by $Gal(E/F)$.

Proposition 1.3.8 (Primitive Element Theorem, [4])
Let $E = F(\alpha_1, \dots, \alpha_n)$ be a finite extension of F. Assume that $\alpha_2, \dots, \alpha_n$ are separable over F. Then there is a $\gamma \in E$ such that $E = F(\gamma)$. Here, $F(\alpha)$ denotes the smallest subfield of $\overline{F}$ that contains both α and F.

PROOF We proceed with induction. Note first that for $n = 1$ there is nothing to prove.

$n = 2$: Let $E = F(\alpha, \beta)$ with β separable over F and let m_α, m_β be their minimal polynomials over F. Let L be a field where $m_\alpha \cdot m_\beta$ splits into linear factors. Obviously, L contains a copy of E. Let $\alpha = \alpha_1, \dots, \alpha_n$ be the roots of m_α and let $\beta = \beta_1, \dots, \beta_m$ be the roots of m_β. Since we are dealing with separable extensions, these roots β_i are distinct. Therefore, the following polynomial equation has exactly one solution for each $i \in \{2, \dots, n\}$ and $j \in \{2, \dots, m\}$:

$$\alpha_i + \beta_j \cdot X = \alpha + X \cdot \beta \tag{1.2}$$

Without loss of generality, let F be infinite. Otherwise, the multiplicative subgroup $E^\star$ is finite and as a subgroup of a field cyclic (compare [4, proposition 4.19], page 53), so it is generated by one element ζ. Due to E being a field this means that $E = F(\zeta)$.
Let us choose a $c \in E$ different from the solutions to (1.3). This surely exists, since we only view a finite amount of solutions but the field is infinite. Set $\gamma = \alpha + c \cdot \beta$. Then the polynomials m_β and $f(x) = m_\beta(\gamma - c \cdot x)$ have coefficients in $F(\gamma)$, since before they had coefficients in F and have β as a root. In fact, β is their only common root, as $\gamma - c \cdot \beta_j \neq \alpha_i$ for $i \neq j$. The greatest common divisor of the polynomials is $x - \beta$ which, in turn, implies that $\beta \in F(\alpha, \beta)$. Then again, the linear combination $\alpha = \gamma - c \cdot \beta$ also lies in $F(\gamma)$, wherefore $F(\gamma) = F(\alpha, \beta)$.

$n \to n+1: \; F(\alpha_1, \alpha_2, \ldots, \alpha_{n+1}) \overset{n=2}{=} F(\alpha_1', \alpha_3, \ldots, \alpha_{n+1}) \overset{ind.}{=} F(\gamma)$ □

Theorem 1.3.9 (Fundamental Theorem of Galois theory for finite extensions, [1]) *Let E be a finite Galois extension of F and let $G = Gal(E/F)$ be the corresponding Galois group. The maps $H \mapsto E^H$ and $M \mapsto Aut(E/M)$ are inverse bijections between the following sets:*

$$\{H : H \subset G \text{ subgroup}\} \longleftrightarrow \{M : E \supset M \supset F \text{ and } M \text{ field}\}$$

Moreover,

1. *The correspondence is inclusion-reversing: $H_1 \supset H_2 \Leftrightarrow E^{H_1} \subset E^{H_2}$.*
2. *The extension E/M is always Galois.*
3. *M/F is galois $\Leftrightarrow$ H is a normal subgroup of G. In this case we have $Gal(M/F) \cong G/H$.*

PROOF For the proof, refer to Szamuely, [1], pages 5f.

Proposition 1.3.10 *The field extension E/F is Galois, if and only if the only elements that are fixed under the action of $Aut(E/F)$ are the ones in F.*

PROOF For the first implication, notice that the action of $Aut(E/F)$ on E indeed fixes F. We need to show that, assuming E/F is a Galois extension, those are the only elements. For that matter, let $\alpha \in E \backslash F$. Without loss of generality, we assume that E is a non-trivial field extension, since for the case E/E the result is trivial. We aim to show that there is $\sigma \in Gal(E/F)$ that does not fix α. Suppose there was no such σ. That means that the action of $Gal(E/F)$ fixes all of $F(\alpha) \subset E$ which, in turn, means that $Gal(E/F(\alpha))$ is trivial. Observe that the extension $E/F(\alpha)$ is Galois by theorem 1.3.9. Since $|Gal(E/F(\alpha))| = [E : F(\alpha)]$ by [1, corollary 1.2.7], page 6, it holds that $F(\alpha) = E$ which is a contradiction.

Then again, assuming that the action of $Aut(E/F)$ only fixes F, we need to check that the field extension E/F is separable and normal. First, pick $\alpha \in E \setminus F$. It is a root of the polynomial $f(x) = \prod_\sigma (x - \sigma(\alpha))$, where σ runs over a system of left coset representatives of

$$G(\alpha) = \{\alpha : g(\alpha) = \alpha \text{ for all } g \in G = Aut(E/F)\}.$$

Its coefficients are in F, since $Aut(E/F)$ fixes only F. As all the roots of f need to be roots of the minimal polynomial m_α, f is finite which is why it is well-defined. Thus, f divides m_α, but since m_α is irreducible by lemma 1.3.5, so is f. Using that $Aut(E/F)$ fixes only F, we see that f has no multiple roots. Therefore, f is a separable polynomial and α is separable, which implies that E/F is a separable field extension.
In like manner, that the field extension is normal, since every irreducible polynomial can be written in the above way modulo factors. Consequently, it splits into linear factors in E, since all the $\sigma(\alpha)$ are contained in E by definition. Finally, this yields that E/F is a Galois field extension.
□

Remark: Here E^G comes from definition 1.2.6 and denotes the set of G-invariants of E.

1.4 The Fundamental Group

Definition 1.4.1 Let X be a topological spaces and let $f,\ g : [0,1] \to x$ be paths in X with the same initial point x_0 and end point x_1. We say that f is *path-homotopic* to g, if there exists a continuous map $H : [0,1]^2 \to X$ such that

$$\begin{array}{lcl} H(s,0) = f(s) & \text{and} & H(s,1) = g(s) \\ H(0,t) = x_0 & \text{and} & H(1,t) = x_1 \end{array}$$

for each $s,\ t \in [0,1]$. We call H a *path-homotopy* between f and g and write $f \ \sim\ g$.

Remark: $\sim$ is an equivalence relation. This is easy to check by giving simple path-homotopies between arbitrary paths. We denote the *path-homotopy class* of a path γ by $[\gamma]$.

Definition 1.4.2 Let γ_0 be a path in a topological space X from x_0 to x_1 and γ_1 a path in X from x_1 to x_2. We define the *product* between γ_0 and γ_1 to be

$$\gamma_0 \star \gamma_1 = \begin{cases} \gamma_0(2s) & \text{for } s \in [0, \frac{1}{2}], \\ \gamma_1(2s-1) & \text{for } s \in (\frac{1}{2}, 1]. \end{cases}$$

This is well-defined and continuous, since $\gamma_0(1) = x_1 = \gamma_1(0)$ and therefore a path from x_0 to x_1.

Lemma 1.4.3 *The operation $\star$ is associative and induces a well-defined operation on path-homotopy classes, if $f(1) = g(0)$:*

$$[f] \star [g] = [f \star g]$$

Furthermore, any path-homotopy class $[\gamma]$ of paths from x_0 to x_1 has a left- and right identity element $[e_{x_0}]$ respectively $[e_{x_1}]$, where e_x is the constant path $e_x(t) = x$ for all $t \in [0,1]$, such that $[e_{x_0}] \star [\gamma] = [\gamma] = [\gamma] \star [e_{x_1}]$ and induces the reverse of γ, denoted by $\overline{\gamma}$, with $\overline{\gamma}(s) = \gamma(1-s)$, such that

$$[\gamma] \star [\overline{\gamma}] = [e_{x_0}] \qquad \text{and} \qquad [\overline{\gamma}] \star [\gamma] = [e_{x_1}]$$

Therefore, one might argue that $[\overline{\gamma}]$ can be called the inverse of $[\gamma]$.

PROOF For the proof, the interested reader is referenced to Munkres' Topology, [2], pages 327ff.

Definition 1.4.4 Let X be a topological space and $x_0 \in X$. A path in X that begins and ends in x_0 is called a *loop* based at x_0. The set of path-homotopy classes based at x_0 together with the operation $\star$ is called the *fundamental group* of X relative to the base point x_0 and we denote it by $\pi_1(X, x_0)$.

Proposition 1.4.5 *The fundamental group $\pi_1(X, x_0)$ is actually a group.*

PROOF By Lemma 1.4.3 we know that the group operation $\star$ is well-defined, since the starting point of any loop is equal to the end-point. In addition, there exists a unique identity element $[e_{x_0}]$, associativity holds and for each class of loops $[\gamma]$ there is a unique inverse element $[\overline{\gamma}]$. □

Definition 1.4.6 Let X and Y be topological spaces, $x_0 \in X$, $y_0 \in Y$ and $h : X \to Y$ a continuous map such that $h(x_0) = y_0$. We define $h_\star : \pi_1(X, x_0) \to \pi_1(Y, y_0)$ by

$$h_\star([\gamma]) = [h \circ \gamma]$$

and call $h_\star$ the homomorphism induced by h.

Theorem 1.4.7 *Let $h : X \to Y$ and $k : Y \to Z$ be continuous maps such that $h(x_0) = y_0$ and $k(y_0) = z_0$ for some $x_0 \in X$, $y_0 \in Y$ and $z_0 \in Z$. Then $(k \circ h) = k_\star \circ h_\star$. If id_X is the identity map on X, then $id_{X\star}$ is the identity homomorphism.*

PROOF By definition 1.4.6 for a path γ in X it holds that

$$(k \circ h)_\star([\gamma]) = [(k \circ h) \circ \gamma] = [k \circ (h \circ \gamma)] = k_\star([h \circ \gamma]) = (k_\star \circ h_\star)([\gamma]).$$

Similarly, we show that $id_{X\star}([\gamma]) = [id_X \circ \gamma] = [\gamma]$ □

Corollary *If $h : X \to Y$ is a homeomorphism, then $h_\star$ is a group-isomorphism of $\pi_1(X, x_0)$ and $\pi_1(Y, y_0)$.*

PROOF Let $g : Y \to X$ be the inverse of h. It then holds that

$$h_\star \circ g_\star = (h \circ g)_\star = id_Y \qquad \text{and} \qquad g_\star \circ h_\star = (g \circ h)_\star = id_X$$

and thus, $h_\star$ is a bijection and consequently a group-isomorphism. □

Remark: The above corollary implies that, in a path-connected space X, the fundamental group $\pi_1(X, x_0)$ is independent of the base point $x_0 \in X$. In such a case, we can speak of the *fundamental group of* X and denote it by $\pi_1(X)$.

1.5 Covering Spaces

Definition 1.5.1 Let E and X be topological spaces and $p : E \to X$ a continuous, surjective map. Suppose for each $x \in X$ there is an open neighborhood $U_x \subset X$ such that $p^{-1}(U_x)$ is a union of some disjoint open $V_\alpha \subset E$, where the restriction $p|_{V_\alpha} : V_\alpha \to U_x$ defines a homeomorphism for each α. Then, we call E a *covering space* of X and p the associated *covering map*. This pair will be referred to as the covering space (E, p).
We call it a *finite covering*, if for each $x \in X$ the preimage $p^{-1}(x)$ contains only finitely many elements.

We can imagine the notion of covering spaces over any of the above described neighborhoods U to be a disjoint union of "topological copies" of U that are all mapped homeomorphically onto U by p. This is displayed in figure 1.1.

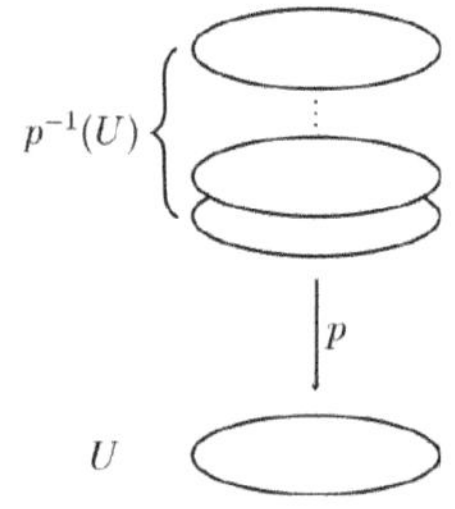

Figure 1.1: Respresentation of an even covering of the open set U [1]

Definition 1.5.2 A path-connected topological space X is called *simply connected*, if any path between two points in X can be continuously transformed into any other between the end points in X.

[1] https://en.wikipedia.org/wiki/Covering_space

Proposition 1.5.3 *Let X be a path-connected topological space. Then X is simply connected if and only if it has trivial fundamental group.*

PROOF Let us first assume that X is simply connected and let $\gamma : x_0 \rightsquigarrow x_0$ be a loop for $x_0 \in X$. Since X is simply-connected, by definition 1.5.2 that means that we can continuously transform γ into any other loop in X, meaning that there is a path-homotopy between γ and any other loop in X. Therefore, the fundamental group of X is trivial.

Next, let us assume X has trivial fundamental group and let $\gamma_0,\ \gamma_1 : x_0 \rightsquigarrow x_1$ be two arbitrary paths in X for some points $x_0,\ x_1 \in X$. They are well-defined, since X is path-connected. Then $\gamma_0 \star \overline{\gamma_1} : x_0 \rightsquigarrow x_0$ is a loop in X and consequently, using definition 1.4.4 and the fact that X is simply connected: $\gamma_0 \star \overline{\gamma_1} \sim e_{x_0}$. This is why

$$[\gamma_0] = [\gamma_0 \star \overline{\gamma_1}] \star [\gamma_1] = [e_{x_0}] \star [\gamma_1] = [\gamma_1],$$

implying X has trivial fundamental group. □

Definition 1.5.4 A *deck transformation* of a cover $p : E \to X$ of a topological space X is a homeomorphism $f : E \to E$ such that $p \circ f = p$.

Remark: The set of all deck transformations on a cover $p : E \to X$ forms a group. We denote this group by $Aut(E/X)$.

Definition 1.5.5 Let $p : E \to B$ be a map and let $f : C \to B$ be a continuous map. A *lifting* of f is a map $\tilde{f} : C \to E$ such that $p \circ \tilde{f} = f$. This is illustrated in the following diagram.

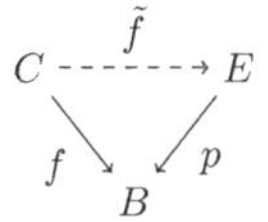

Lemma 1.5.6 (The lifting Criterion, [9])
Let $p : E \to B$ be a covering. Pick $e_0 \in E$ and fix $b_0 = p(e_0)$. Let C be a path-connected and locally path-connected topological space and let $f : C \to B$ be a continuous map and $c_0 \in C$ such that $f(c_0) = b_0$. Then a continuous lift $\tilde{f} : C \to E$ of f such that $\tilde{f}(c_0) = e_0$ exists and is unique, if and only if

$$f_\star(\pi_1(C, c_0)) \subset p_\star(\pi_1(E, e_0))$$

where $f_\star$ and $p_\star$ are induced homomorphisms from definition 1.4.6.

PROOF First if such a lift $\tilde{f}$ exists, we know from definition 1.5.5 that $f_\star = p_\star \circ \tilde{f}_\star$. Therefore,

$$f_\star(\pi_1(C, c_0)) = p_\star(\tilde{f}_\star(\pi_1(C, c_0))) \subset p_\star(\pi_1(E, e_0))$$

as the image of a map is always contained in its domain and we know $\tilde{f}_\star : \pi_1(E, e_0) \to \pi_1(C, c_0)$.

For the other implication and uniqueness, see Glickenstein [9], propositions 7 and 8 on pages 2f.

Corollary *Let $p : E \to B$ be a cover and $e_0 \in E$. We set $b_0 = p(e_0)$. Let $\gamma_1, \gamma_2 : [0,1] \to B$ be paths with $\gamma_1(0) = \gamma_2(0) = b_0$ and $\gamma_1(1) = \gamma_2(1) = b_1$ for some $b_1 \in B$. Then, for every homotopy $H : [0,1] \times [0,1] \to B$ between γ_1 and γ_2 there is a unique lifting homotopy $\tilde{H} : [0,1] \times [0,1] \to E$ such that $\tilde{H}(0,0) = e_0$. Since $\tilde{H}$ is a lift, we also have that $p(\tilde{H}) = H$.*

PROOF We observe that the rectangle $[0,1] \times [0,1]$ has trivial fundamental group. Hence, the condition of lemma 1.5.6 is fulfilled and we can lift H to a unique homotopy $\tilde{H} : [0,1] \times [0,1] \to Y$ such that $\tilde{H}(0,0) = e_0$ and $p \circ \tilde{H} = H$. This property specifically gives us that $p \circ \tilde{H}(t, 0) = \gamma_1(t)$ and $p \circ \tilde{H}(t, 1) = \gamma_2(t)$, so we define our paths $\tilde{\gamma_1}$ and $\tilde{\gamma_2}$ accordingly. □

Chapter 2

Galois Categories

2.1 Definition

Definition 2.1.1 (Axioms of a Galois Category, [3] and [5])
Let $\mathcal{C}$ be a category and $F : \mathcal{C} \to sets$ a covariant functor from $\mathcal{C}$ to the category of finite sets. We call $\mathcal{C}$ a *Galois Category* with *fundamental functor* F, if the following conditions are met:

G1. There is a *final object* in $\mathcal{C}$ and the *fiber product* of any two objects over a third exists in $\mathcal{C}$.

G2. $\mathcal{C}$ has an *initial object*, *finite coproducts* exist in $\mathcal{C}$, and the quotient of an object by a finite group of automorphisms exists in $\mathcal{C}$.

G3. Any morphism $u \in Hom(\mathcal{C})$ can be written $u = u' \circ u''$ where u'' is an *epimorphism* and u' is a *monomorphism* and any *monomorphism* $u \in Hom_{\mathcal{C}}(X, Y)$ is an isomorphism of X with a direct summand of Y.

G4. The functor F maps final objects to final objects and commutes with fiber products.

G5. F commutes with finite coproducts and quotients (as above) and maps epimorphisms to epimorphisms.

G6. If $u \in Hom(\mathcal{C})$ and $F(u)$ is an isomorphism, then u is an isomorphism.

Explanation: Even though we have defined all of the terms in the chapter of algebraic foundations, this definition appears to be quite long. In the following, all steps will be clarified to understand specifically what each point means in the sense that the target is the set of finite sets. For better understanding, let us prove that the category of finite sets itself, together with the identity functor $F = id_{\mathcal{C}}$ is a Galois Category by following Lynn[3], page 10:

G1. First of all, the final object Z by definition 1.1.3 satisfies that for each object X there is a unique morphism $f : X \to Z$. Here, this has to be the singletons $\{x\}$, since the set of maps from a set A into a set S has $|S|^{|A|}$ elements. Thus, $|S| = 1$ is the only option. We note that this implies that final objects are unique up to a canonical isomorphism.
We define the fiber product from definition 1.1.10 of two sets A, B over a set Z to be $A \times_Z B = \{(x, y) \in X \times Y : f(x) = g(y)\}$ with $f : X \to Z,\ g : Y \to Z$ derived from the definition of a pullback. This exists for any two sets A, B.

G2. The initial object A by definition 1.1.3 is for any object X there is a unique morphism $f : A \to X$. However, the set of maps from A to X has cardinality $|X|^{|A|}$, which should be equal to 1, and the morphism is supposed to be unique, so we find out that $|A| = 0$ and therefore $A = \emptyset$.

We define the finite coproduct of sets X_i, following definition 1.1.12, to be the usual disjoint union $\bigsqcup_{i\in I} X_i$ and following definition 1.1.13 the quotient to be

$$X/G = \{G \circ x : x \in X\} = \{\ \{g \circ x : g \in G\}\ : x \in X\}.$$

G3. Epimorphisms between sets are basically surjective maps, while monomorphisms between sets are injective functions, so let $u : B \to D$ be any map. Suppose some elements are mapped to the same element by u and fix an equivalence relation on B with $x \sim y \Leftrightarrow u(x) = u(y)$. Then the quotient map $u'' = p : B \to B/\sim$ is surjective and because we factored out elements that mapped to the same element, the restriction $u' = u\big|_{B/\sim}$ is injective.
Now, let $u : X \to Y$ be a monomorphism (injective function). Following definition 1.1.15, we pick $Z = Y \setminus u(X)$. The inclusion map $\iota : Z \to Y$ together with u induce the disjoint union $Y = u(X) \sqcup Y \setminus u(X)$, which makes Y a coproduct of X and Z together with u and ι and yields that any monomorphism $u \in Hom_{\mathcal{C}}(X, Y)$ is an isomorphism of X with a direct summand of Y.

G4. Since F is the identity functor $sets \to sets$, obviously final objects are mapped to final objects and $F(X \times_Z Y) = id_{\mathcal{C}}(\{(x, y) : f(x) = g(y)\}) = X \times_Z Y = F(X) \times_{F(Z)} F(Y)$.

G5. For the finite coproduct we get $F(X \sqcup Y) = X \sqcup Y = F(X) \sqcup F(Y)$. Since an epimorphism by the identity functor is mapped to itself, it stays an epimorphism. The quotient yields:

$$F(X/G) = \{\{g \circ x : g \in G\}\ : x \in X\} = \{\{g \circ x : g \in F(G)\}\ : x \in F(X)\} = F(X)/F(G)$$

G6. F maps isomorphisms to itself and therefore G6 also holds.

As we have shown that all axioms of a Galois category hold, this implies that the set of finite sets *sets* is itself a Galois category with fundamental functor id_{sets}.

Definition 2.1.2 A category $\mathcal{C}$ is called *essentially small*, if it is equivalent to a category, whose objects form a set.

Proposition 2.1.3 *Any essentially small Galois category is equivalent to the category of finite π-sets for a uniquely determined profinite group π.*

Proof Our immediate action will be to recall that a π-set by definition 1.2.6 is nothing but a set, on which π acts and that an essentially small category was defined in 2.1.2. This proposition is proven in Lentra, [5, sections 3.3 - 3.5 and 3.11 - 3.19], pages 39f., in great detail and would go beyond the scope of this thesis. □

2.2 Infinite Galois Theory

Remember that an algebraic field extension E/F is called an *Galois extension*, if it is both separable and normal, both terms derived from definition 1.3.6. This, on the other hand, means that each irreducible polynomial $f \in F[X]$ having a root in E has $deg(f)$ distinct roots in E.
In section 1.3 we assumed the field extensions to be finite but we can now easily extend the above definition for Galois extensions to infinite field extensions, opening up a whole new field of study. This brings us to the following definition:

Definition 2.2.1 Let k be some field. Its *algebraic closure* $\overline{k}$ is an algebraic field extension such that

$$\overline{k} = \{\alpha : f(\alpha) = 0 \text{ for some } f \in k[X]\}.$$

Its *separable closure* k^{sep} is a subfield of $\overline{k}$ and can be defined as

$$k^{sep} = \{\alpha \in \overline{k} : m_\alpha \in k[X] \text{ has distinct roots in } \overline{k}\}$$

where m_α is the minimal polynomial of α from definition 1.3.4. It is also a field extension of k.

Let $\alpha \in k^{sep}$ and let $m_\alpha \in k[X]$ be its minimal polynomial from definition 1.3.4. Since $\overline{k}$ is normal over k by the above definition, m_α splits in $\overline{k}$. We write

$$m_\alpha(x) = \prod_{i=1}^{n}(x - a_i)$$

By the above definition, we know that all roots are distinct, since $\alpha \in k^{sep}$. Since they have the same minimal polynomial, they necessarily also have to be in k^{sep} which implies that k^{sep} is normal over k and therefore Galois. Therefore the below definition is justified.

Definition 2.2.2 The Galois group of the field extension k^{sep}/k is called the *absolute Galois Group* of k and is denoted by $Gal(k) = Gal(k^{sep}/k)$.

Lemma 2.2.3 *If Ω is a Galois extension of F, it is Galois over every intermediate field M.*

PROOF We already proved this for finite Galois extensions in 1.3.9. Now, assume the inclusions $F \subset M$ and $M \subset \Omega$ are proper and let $f \in M[X]$ be an irreducible polynomial with a root α_1 in Ω. This means that $f(x) = (x - \alpha_1) \cdot g(x)$, where $g(x)$ is irreducible over $M[X]$. Suppose $g(\alpha_1) = 0$, then α_1 would be a double zero in $M^{sep} \supset F^{sep} \supset E$ and therefore M^{sep} is not Galois over M which is a contradiction. Therefore all roots of f are distinct. Now, let $\alpha_1, \dots, \alpha_n$ be all roots of f and let $m_{\alpha_1}, \dots, m_{\alpha_n}$ be their minimal polynomials over F. Since E is Galois over F, the minimal polynomials split in linear factors in E. Consequently, $f(x) \mid \prod_{i=1}^{n} m_{\alpha_i}$ in $E[X]$ which, since the roots of f are distinct, means that f also splits in linear factors in $E[X]$. □

Proposition 2.2.4 *Let k be a field, k^{sep} its separable closure and $E \subset k^{sep}$ a subfield containing k. The following are equivalent:*

1. *The extension E/k is Galois.*
2. *For each automorphism $\sigma \in Gal(k)$ it holds that $\sigma(E) \subset E$.*

PROOF We begin with the proof of (2) $\Rightarrow$ (1). As we know from the notion above, k^{sep}/k is a Galois extension. Let us assume that the inclusions from the proposition are proper, since otherwise the proof would be trivial. Let us now pick some $\alpha \in E \setminus k$. Since the field extensions are non-trivial, we find a non-trivial element $\sigma \in Gal(k)$ that satisfies $\sigma(\alpha) \neq \alpha$. This exists due to $(k^{sep})^{Gal(k)} = k$. The equation holds because we can apply lemma 2.2.3 in the proof of proposition 1.3.10 to get a more general statement.

By our assumption, $\sigma(E) \subset E$ and thus we have for the restriction $\sigma\big|_E \in Aut(E/k)$ that it does not fix α. Therefore the elements that remain fixed under the action of $Aut(E/k)$ (definition 1.2.6) are exactly the ones in k which is equivalent to E/k being Galois by proposition 1.3.10.

Vice versa, for (1) $\Rightarrow$ (2), let $\sigma \in Gal(k)$ be an automorphisms and let $\alpha \in E \setminus k$ be some element, where m_α denotes its minimal polynomial. Since by assumption E/k is Galois, E must by definition 1.3.7 contain all the other roots of m_α. In turn, for being a k-automorphism, σ must permute the roots of m_α. Thus, for all elements α of E we have that $\sigma(\alpha) \in E$, which implies that $\sigma(E) \subset E$. □

In the following, our aim will be to construct for any Galois field extension Ω/k a Galois group that will have similar properties to the one we already dealt with in 1.3. We will use profinite groups to equip the Galois Group with a topology that will be called the *Krull topology*.

Now, let Ω/k be a (possibly infinite) Galois extension, where $Gal(\Omega/k)$ denotes its Galois Group. Let us pick the Galois subextensions of finite degree and let us pick group homomorphisms $\phi_{ML} : Gal(M/k) \rightarrow Gal(L/k)$, if M is a field extension of L that is canonically induced by restriction. This gives us the directed set of all finite Galois field extensions of a field k. Axiom 4 of definition 1.2.2 is met here, since by the theorem of the primitive element 1.3.8 we can write two field extensions L and M of k as $L = k(\gamma)$ and $M = k(\theta)$ for some $\gamma, \theta \in \overline{k}$, since L and M are separable, because they are Galois. Then $k(\gamma, \theta)$ contains both M and L as subfields and still is a Galois extension of finite degree, as both L and M are Galois, applying lemma 1.3.2. By the fundamental theorem of Galois theory 1.3.9 inclusion-reversing correspondence can be given for the Galois groups of these fields. Thus, by definition 1.2.3, the group homomorphisms, the Galois groups of finite field extensions together with inclusion form a projective system. Its projective limit forms a group by proposition 1.2.4 and we denote it by $\lim\limits_{\leftarrow} Gal(L/k)$.

Proposition 2.2.5 *The projective limit* $\lim\limits_{\leftarrow} Gal(L/k)$ *is isomorphic to* $Gal(\Omega/k)$.

Proof First, let us define a group homomorphism $\psi : Gal(\Omega/k) \rightarrow \prod_{[L:k]<\infty} Gal(L/k)$ by mapping $\sigma \mapsto \sigma\big|_{L_1} \times \sigma\big|_{L_2} \times \ldots$ for finite Galois extensions L_i. Proposition 2.2.4 gives us the fact that $\sigma(L) \subset L$ for any of the finite Galois extensions, because every separable extension is contained in k^{sep}. Consequently, ψ is well-defined. This map is injective, since if σ does not fix $\alpha \in \Omega$, then $\sigma\big|_L$ is not trivial for some finite field extension L of k that contains $k(\alpha)$ and by the theorem of the primitive element 1.3.8 every subextension of finite degree can be written that way. Thus, the kernel only contains the identity. By the fundamental theorem of Galois theory 1.3.9, the image of ψ is contained in the projective limit $\lim\limits_{\leftarrow} Gal(L/k)$ that, by definition 1.2.3, is a subset of the domain.
Let us now pick $(\sigma_L)_{[L:k]<\infty} \in \lim\limits_{\leftarrow} Gal(L/k)$. We define an automorphism σ of Ω that fixes k: For any $\alpha \in \Omega$ set $\sigma(\alpha) = \sigma_L(\alpha)$ for some finite subextension that contains $k(\alpha)$. This is well-defined, since the field extensions are compatible with each other by the above construction. Again, by construction, this exactly maps to $(\sigma_L)_{[L:k]<\infty}$ and therefore, ψ is surjective, implying that it is an isomorphism. □

Let us now leave the point of view of only picking subextensions of some fixed Galois extension. The following corollary applies proposition 2.2.5 by building the projective limit over *all* finite Galois extensions of k. All of them are necessarily contained in k^{sep} and k^{sep} is a Galois extension.

Corollary *The projective system that includes all the finite Galois extensions of a field* k *has a projective limit that is isomorphic to the absolute Galois group* $Gal(k) = Gal(k_{sep}/k)$.

Theorem 2.2.6 (Krull's Fundamental Theorem of Galois Theory, [4])
Let Ω *be a Galois extension of* k *and let* L *be a subextension. Then* $Gal(\Omega/L)$ *is a closed subgroup of* $Gal(\Omega/k)$. *Moreover, the maps* $L \mapsto Gal(\Omega/L)$ *and* $H \mapsto \Omega^H$ *yield an inclusion-reversing bijection between the sets*

$$\{L \text{ field} : \Omega \supset L \supset k\} \longleftrightarrow \{H : H \text{ closed subgroup in } Gal(\Omega/k)\}$$

Furthermore, a closed subgroup H *is normal if and only if* Ω^H *is Galois over* k *and in this case there is a natural isomorphism* $Gal(\Omega^H/k) \cong Gal(\Omega/k) \;/\; H$.

Proof First, let us show the maps given in the statement are inverse maps. For that matter, let H be a closed subgroup of $Gal(\Omega/k)$. By lemma 2.2.3 we get that Ω is Galois over Ω^H, as it is an intermediate field. Then we get by[4, proposition 7.9], page 94, that $Gal(\Omega/\Omega^H) = H$.
Otherwise, let L be a subextension. $Gal(\Omega/L)$ is a profinite group itself, but in this case only using finite Galois field extensions of L that are by lemma 1.3.2 also finite field extensions of k. Therefore, it is a subgroup of $Gal(\Omega/k)$. Then, by the version of proposition 1.3.10 for

infinite Galois extensions that is proved by applying lemma 2.2.3, we know that $\Omega^{Gal(\Omega/L)} = L$. Furthermore, the group $Gal(\Omega/L)$ is closed in $Gal(\Omega, k)$ by [4, proposition 7.11(a)], pages 94f.

The proof of the last statement can be found in Szamuely, [1], pages 95f. □

Example: Notice the difference to the fundamental theorem of finite Galois theory: In this case, not every subgroup of $Gal(E/F)$ is considered, but only the closed ones. If we consider the groups $\mathbb{Z}/p\mathbb{Z}$ for primes $p \geq 2$, using example [$\hat{\mathbb{Z}}$] and proposition 2.2.5, its absolute Galois group is $\hat{\mathbb{Z}}$. This implies, again by example [$\hat{\mathbb{Z}}$], that it contains a non-closed subgroup, namely a copy of $\mathbb{Z}$, proving that not all subgroups of the Galois group are necessarily closed.

2.3 Finite Étale Algebras

Definition 2.3.1 Let k be a field and let A be a k-*algebra*, meaning that A is a ring with 1 together with a ring homomorphism $\psi_A : k \to A$ such that $\psi_A(1_k) = 1_A$ and for all $x, y \in im(\psi_A)$ it holds that $xy = yx$.
The k-algebra A is called *étale*, if it is isomorphic to a finite product of separable extensions of k. When all these extensions have finite degree over k, it is said to be *finite étale*.

Now, let ${}_kSAlg$ denote the category of finite étale algebras. The morphisms between two objects $A, B \in ob({}_kSAlg)$ are k-algebra homomorphisms $\phi : A \to B$, i.e. a ring homomorphisms such that for all $x \in k$ and $a \in A$ it holds that $\phi(xa) = x\phi(a)$.
Then, let us denote the opposite category of ${}_kSAlg$ by $\mathcal{C}_k = {}_kSAlg^{op}$. In the following, we will show that this is a Galois category together with the fundamental functor $F_k = Hom_{{}_kSAlg}(-, \overline{k})$, where $\overline{k}$ denotes a fixed algebraic closure. This sends a k-algebra A to the k-algebra homomorphisms from A into $\overline{k}$. Since we only deal with finite étale algebras, this set is clearly finite.
Let us now prove that the category $\mathcal{C}_k$ is a Galois category. However, before doing that, we need to introduce some lemmas.

Lemma 2.3.2 *The functor* $- \otimes_k B$ *for a* k*-algebra* B *preserves and detects epimorphisms and monomorphisms.*

Proof (Sketch) A k-algebra B over a field k is a k-vector space equipped with a bilinear product. For an epimorphism $u : A \to B$ it suffices to choose a basis $v_1, \dots, v_n, v_{n+1}, \dots, v_m$ in A and choose from the image of these m vectors a basis $(w_i)_{i=1}^n$ for B, which is possible, because u is a surjective ring homomorphism. Then we apply this construction to the map $u : A \otimes_k C \to B \otimes_k C$ for a k-algebra C and get that for a basis $(c_j)_{j \in J}$ for some J of C we have that $u(v_i \otimes c_j) = w_i \otimes c_j$ for $i \in [n]$ and j arbitrary. But this already generates $B \otimes_k C$, since $\{w_i \otimes c_j : i \in [n],\ j \in J\}$ is a basis of it, so this map is an epimorphism.
The other proofs are analogous to the above. □

Lemma 2.3.3 *Let* k *be a field and* $\overline{k}$ *an algebraic closure of* k*. Let* I *be a finite set. Then*

$$Hom_{\overline{k}}(\textstyle\prod_I \overline{k}, \overline{k}) \cong I$$

Proof This isomorphism sends $i \in I$ to $(a_1, \dots, a_n) \mapsto a_i$ in $Hom_{\overline{k}}(\prod_I \overline{k}, \overline{k})$.
Its inverse map is for any $\phi \in Hom_{\overline{k}}(\prod_I \overline{k}, \overline{k})$ given by the map we will construct in the following. To do that, we classify the morphisms in $Hom_{\overline{k}}(\prod_I \overline{k}, \overline{k})$. First, we notice that the image of every $\overline{k}$ is determined by the image of $e_i = (0, \dots, 0, 1, 0, \dots, 0)$, since $\phi(ax) = a\phi(x)$ for all $a \in \overline{k}$ and $x \in \prod_I \overline{k}$. Let us now assume that some e_i and e_j for $i \neq j$ are mapped to a non-zero element. Then $0 = \phi((0, \dots, 0)) = \phi(e_i \cdot e_j) = \phi(e_i) \cdot \phi(e_j)$ with the component-wise multiplication in $\prod_I \overline{k}$. The fact that $\overline{k}$ is a field implies that it contains no zero-divisors, which yields that either $\phi(e_i)$ or $\phi(e_j)$ are zero. This is a contradiction, so at most one entry in $\prod_I \overline{k}$ can be mapped to

non-trivial elements.
However, ϕ fixes $\overline{k}$, since it is a $\overline{k}$-algebra homomorphism, so it can't be the zero-map, since $\{0\}$ is no field. This implies that all morphisms in $Hom_{\overline{k}}(\prod_I \overline{k}, \overline{k})$ are given by projections in the i-th coordinate. Subsequently, the inverse map is given by $\pi_i \mapsto i$. □

Lemma 2.3.4 *Let A be a k-algebra for a field k. Then $A \otimes_k \overline{k} \cong \prod_I \overline{k}$ for some finite set I, if and only if A is finite étale.*

PROOF For the proof see Lenstra [5, proposition 2.7(iii)⇔(iv)], pages 21f. □

Finally, this gives us the necessary means to state the following proposition.

Proposition 2.3.5 *The category $\mathcal{C}_k$ together with the fundamental functor $F_k = Hom_{{}_kSAlg}(-, \overline{k})$ is a Galois category.*

PROOF Remember that by definition 2.1.1, the category $\mathcal{C}_k$ needs to fulfill 6 axioms to be a Galois category. We use the definition of ${}_kSAlg$ and F_k from above to prove them.

G1. The final object in $\mathcal{C}_k$ is the initial object in ${}_kSAlg$ which is k. It is initial in ${}_kSAlg$, since for any object $A \in ob({}_kSAlg)$ there is precisely one morphism $\phi : k \to A$, as we assume that $\phi(x) = x\phi(1_k) = x \cdot 1_A$ for all $x \in k$. Thus, k uniquely determines every k-algebra homomorphism with domain k.

We define the fiber product in $\mathcal{C}_k$ to be the coproduct in ${}_kSAlg$, the tensor product: For any two k-algebras $A, B, C \in ob(\mathcal{C}_k)$ it holds that $A \times_C B = A \otimes_C B$.

G2. The initial object in $\mathcal{C}_k$ is the final object in ${}_kSAlg$. This is the null-ring $\{0\}$. Indeed, there is a single morphism from any k-algebra A into $\{0\}$ (the 0-homomorphism). Moreover, $\{0\}$ is a separable k-algebra, since it satisfies the definition, as it is the empty product.

The finite coproduct $A \sqcup B$ of two objects $A, B \in ob(\mathcal{C}_k)$ will be defined to be the k-algebra generated by A and B such that if $A \cong k_1 \times \cdots \times k_n$ and $B \cong k_{n+1} \times \cdots \times k_m$, we have $A \sqcup B \cong k_1 \times \cdots \times k_m$. This is the product in ${}_kSAlg$.

The categorial quotient A/G for a group of automorphisms G in $\mathcal{C}_k$ is the set of G-invariants of A in ${}_kSAlg$, namely

$$A^G = \{a \in A : g(a) = a \text{ for all } g \in G\}.$$

G3. Let $u : A \to B$ be a morphism in $\mathcal{C}_k$. It factors to the epimorphism $u' : A \to im(u)$ with $a \mapsto u(a)$ and the monomorphism $u'' : im(u) \to B$ with $b \mapsto b$. Then we consider $u = u'' \circ u'$.

What remains to be checked is that $im(u) \cong A/I$ for the ideal $I = ker(u)$ is indeed a finite étale algebra. We notice that A/I is a ring and also a k-algebra by restricting the map ψ_A. Since $u' : A \to A/I$ is an epimorphism, the map $u' : A \otimes_k \overline{k} \to (A/I) \otimes_k \overline{k}$ is also an epimorphism by lemma 2.3.2. By lemma 2.3.4, $(A/I) \otimes_k \overline{k} \cong (\prod_I \overline{k})/J$ for some ideal J. We claim that every ideal J of $\prod_I \overline{k}$ is of the form $\prod_{I'} \overline{k}$ for some $I' \subset I$. Indeed, for some $(a_i)_{i \in I'} \in J$ with $a_n \neq 0$ for some n we have that $e_i = (0, \ldots, 0, 1, 0, \ldots, 0) \in J$. For that, we simply choose $(0, \ldots, a_n^{-1}, \ldots 0)$ and multiply it with $(a_i)_{i \in I'}$. The existence of a_n^{-1} is guaranteed, since $\overline{k}$ is a field.

This implies that $J = \prod_{I'} \overline{k}$ where $I' = \bigcup_{a \in J} \{i \in I : a_i \neq 0\}$. Now clearly, this yields that $A/I \otimes_k \overline{k} \cong \prod_I \overline{k} / \prod_{I'} \overline{k} \cong \prod_{I \setminus I'} \overline{k}$. Subsequently, applying lemma 2.3.4, this means that A/I is a finite étale algebra.

We omit the rest of the proof for the sake of brevity.

G4. Knowing that $F_k(k) = Hom_k(k, \overline{k})$, as noted in G1 this set consits only of one element, F_k maps final objects to final objects.

Furthermore, for the tensor product of some $A, B \in ob(\mathcal{C}_k)$ over $C \in ob(\mathcal{C}_k)$ it holds that:

$$\begin{aligned}
F_k(A \otimes_C B) &= Hom_k(A \otimes_C B, \overline{k}) \\
&= \{C\text{-bilinear maps } \Psi : A \times B \to \overline{k}\} \\
&= \{\Psi : A \times B \to \overline{k} : \Psi(ca, b) = c\Psi(a, b) = \Psi(a, cb)\} \\
&= \{(\alpha \times \beta) \in \left(Hom_k(A, \overline{k}) \times Hom_k(B, \overline{k})\right) : F_k(\psi_A)(\alpha) = F_k(\psi_B)(\beta)\} \\
&= Hom_k(A, \overline{k}) \times_{Hom_k(C,\overline{k})} Hom_k(B, \overline{k}) = F_k(A) \times_{F_k(C)} F_k(A)
\end{aligned}$$

which implies that F_k commutes with the fiber product.

G5. We view a coproduct of two objects. Induction yields the result for greater numbers.

$$\begin{aligned}
F_k(A \times B) &= Hom_k(A \times B, \overline{k}) \\
&= Hom_{\overline{k}}((A \times B) \otimes_k \overline{k}, \overline{k}) \\
&= Hom_{\overline{k}}((A \otimes_k \overline{k}) \times (B \otimes_k \overline{k}), \overline{k}) \\
&\overset{2.3.4}{=} Hom_{\overline{k}}(\prod_I \overline{k} \times \prod_J \overline{k}, \overline{k}) \\
&= Hom_{\overline{k}}(\prod_{I \sqcup J} \overline{k}, \overline{k}) \\
&\overset{2.3.3}{=} Hom_{\overline{k}}(\prod_I \overline{k}, \overline{k}) \sqcup Hom_{\overline{k}}(\prod_J \overline{k}, \overline{k}) \\
&\overset{2.3.4}{=} Hom_k(A, \overline{k}) \sqcup Hom_k(B, \overline{k}) = F_k(A) \sqcup F_k(B)
\end{aligned}$$

Consequently, F_k commutes with finite coproducts.

The fundamental functor $\Phi_{alg} : {}_kSAlg \to sets$ that corresponds to ${}_kSAlg$ maps some arbitrary morphism $u : E \to D$ to a morphism from $Hom_k(D, \overline{k})$ to $Hom_k(E, \overline{k})$ by taking a k-algebra homomorphism $g : D \to \overline{k}$ to $\Phi_{alg}(u)(g) = g \circ u$.
Now, let $u : A \to B$ be an epimorphism in $\mathcal{C}_k$. According to lemma 2.3.2, $u : A \otimes_k \overline{k} \to B \otimes_k \overline{k}$ is an epimorphism. Using lemma 2.3.4, $u : \prod_I \overline{k} \to \prod_J \overline{k}$ is an epimorphism which yields I to have greater cardinality than J. Consequently, we view $F_k(u) : Hom_k(\prod_I \overline{k}, \overline{k}) \to Hom_k(\prod_J \overline{k}, \overline{k})$. This is also an epimorphism, because I has greater cardinality than J and we can apply lemma 2.3.3.

Finally, let $A \in ob(\mathcal{C}_k)$ be a finite étale algebra and G a group of automorphisms. Then:

$$\begin{aligned}
F_k(A/G) &= Hom_k(A^G, \overline{k}) \\
&= Hom_{\overline{k}}(A^G \otimes_k \overline{k}, \overline{k}) \\
&\overset{(-)}{=} Hom_{\overline{k}}((A \otimes_k \overline{k})^G, \overline{k}) \\
&\overset{2.3.4}{=} Hom_{\overline{k}}((\prod_I \overline{k})^G, \overline{k}) \\
&= Hom_{\overline{k}}(\prod_{I/F_k(G)} \overline{k}, \overline{k}) \\
&\overset{2.3.3}{=} Hom_{\overline{k}}(\prod_I \overline{k}, \overline{k})/F_k(G) \\
&\overset{2.3.4}{=} Hom_{\overline{k}}(A \otimes_k \overline{k}, \overline{k})/F_k(G) \\
&= Hom_k(A, \overline{k})/F_k(G) = F_k(A)/F_k(G)
\end{aligned}$$

Here, $(-)$ uses the fact that $A^G \otimes_k \overline{k} = (A \otimes_k \overline{k})^G$. This is clear, after we choose a vector space basis for A^G, extend it to a basis for A, and consider the induced basis of $A \otimes_k \overline{k}$.

G6. Let $u : A \to B$ be a morphism such that $F_k(u) : Hom_k(A, \overline{k}) \to Hom_k(B, \overline{k})$ is an isomorphism. Then, by the definition of the tensor product, we have that

$$F_k(u) : Hom_{\overline{k}}(A \otimes_k \overline{k}, \overline{k}) \to Hom_{\overline{k}}(B \otimes_k \overline{k}, \overline{k})$$

is also an isomorphism. This indicates that $F_k(u) : Hom_{\overline{k}}(\prod_I \overline{k}, \overline{k}) \to Hom_{\overline{k}}(\prod_J \overline{k}, \overline{k})$ is an isomorphism by lemma 2.3.4. Thus, by lemma 2.3.3, $I \cong J$.

This gives us that the morphism $u : \prod_I \overline{k} \to \prod_J \overline{k}$ is an isomorphism. By lemma 2.3.4, $u : A \otimes_k \overline{k} \to B \otimes_k \overline{k}$ is an isomorphism, which implies that $u : A \to B$ is an isomorphism by lemma 2.3.2.

This completes the proof and hence $\mathcal{C}_k$ is a Galois category. □

Theorem 2.3.6 *Let k be a field with fixed separable closure k^{sep}. The functor Φ_{alg} gives us an anti-equivalence between the category of finite étale algebras ${}_kSAlg$ and finite sets with a continuous left $Gal(k)$-action. Here, the sets are $Hom_k(-, \overline{k})$.*

PROOF First, let us note that the category ${}_kSAlg$ is essentially small. Since k^{sep} is a set, so is $T = \prod_{\mathbb{N}} k^{sep}$. This implies that the power set $\mathcal{P}(T)$ of T is also a set and since $ob({}_kSAlg) \subset \mathcal{P}(T)$ by definition 2.3.1, $ob({}_kSAlg)$ forms a set.

Therefore, we can apply proposition 2.1.3 which makes the category $\mathcal{C}_k$ that now is an essentially small Galois category by proposition 2.3.5. This gives us that ${}_kSAlg$ is anti-equivalent to the category π-sets for some profinite group π. For calculating π we can use a construction from Lenstra, [5], pages 35f., that says that π is isomorphic to $Aut(F_k) = Aut(\Phi_{alg})$.

However, by Lenstra [5, section 3.18], pages 42f., we know that $Aut(\Phi_{alg}) \cong \varprojlim Aut_{\mathcal{C}_k}(A)$, where A ranges over the *connected Galois objects* in $\mathcal{C}_k$. For a connected Galois object A we know that $A^{Aut_{\mathcal{C}_k}(A)} = k$, because k is the final object in $\mathcal{C}_k$. Due to theorem 2.2.6 and the fact that $Aut_{\mathcal{C}_k}(A) = Aut(A/k)$, we get that A/k is a Galois extension. This, in turn, implies that

$$\pi \cong Aut(\Phi_{alg}) \cong \varprojlim_{A \text{ conn. Gal.}} Aut_{\mathcal{C}_k}(A) \cong \varprojlim_{L/K \text{ Galois}} Gal(L/k) \overset{2.2.5}{\cong} Gal(k).$$

We will now show that the canonical left action of $Gal(k)$ on $Hom_k(A, k^{sep})$ is continuous. A group action of a profinite group π on a finite set is continuous if and only if the stabiliser of every element is a closed subgroup of π. We also know that some $\sigma \in Gal(k)$ stabilises a k-algebra homomorphism $\phi \in Hom_k(A, \overline{k})$ if and only if it stabilises its image. By definition 2.3.1, A is a product of separable fields, so the image of ϕ is the image of one of the fields. This is already verified above. Subsequently, it is sufficient to show that the stabiliser in $Gal(k)$ of any finite sub-extension of k is closed. However, this was already proved in theorem 2.2.6, so the action of $Gal(k)$ on $Hom_k(A, k^{sep})$ is continuous. □

Chapter 3

Covering Spaces

Following chapter 2 and the ideas of section 1.5, let us at the start of this chapter conclude that the category of finite coverings of a connected topological space X, namely $\mathcal{C}_X$, is a Galois category. Here, the the morphisms $f : Y_1 \to Y_2$ between covers of X, referred to as $p_i : Y_i \to X,\ i \in \{1, 2\}$, satisfy that f is continuous and $p_1 = p_2 \circ f$. The corresponding fundamental functor is

$$F_x : \mathcal{C}_X \to \ sets, \quad F_x(p : Y \to X) = p^{-1}(\{x\})$$

for some fixed $x \in X$. This should give us an idea of how the category of finite coverings behaves, an introduction to its properties and help for further understanding Galois categories. For that reason, let us prove the claim.

Claim: The category $\mathcal{C}_X$ of finite coverings of a connected space X is a Galois Category.

PROOF Recalling definition 2.1.1, we need to check six axioms for Galois categories:

G1. The trivial covering $id : X \to X$ here is a final object, as the morphisms to this object are by the property of the identity map equal to the covering map of the object from the domain. We define the fiber product to be

$$Y_1 \times_Z Y_2 = \{(y_1, y_2) : f(y_1) = g(y_2)\}$$

for morphisms $f : Y_1 \to Z$ and $g : Y_2 \to Z$ with the subspace topology. We still need to check that this is a covering of X together with the obvious restricted covering map $p_1 \times p_2 : Y_1 \times_Z Y_2 \to X$. Let $U \subset X$ be an open set. Then

$$\begin{aligned} (p_1 \times p_2)^{-1}(U) &= (p_1^{-1}(U) \times p_2^{-1}(U)) \ \cap \ Y_1 \times_Z Y_2 \\ &\overset{1.5.1}{=} \left(\bigcup_{\alpha \in I} V_\alpha \times \bigcup_{\beta \in J} W_\beta \right) \ \cap \ Y_1 \times_Z Y_2 \\ &= \bigcup_{\alpha \in I,\ \beta \in J} (V_\alpha \times W_\beta \cap Y_1 \times_Z Y_2), \end{aligned}$$

where the last part in the paranthesis is open in the subspace topology within the product topology and with the fact that restrictions and cartesian products are continuous operations, this yields that $p_1 \times p_2 : Y_1 \times_Z Y_2 \to X$ is indeed a cover.

G2. As the initial object, we can take the empty covering $p : \emptyset \to X$, which obviously satisfies that there is a unique morphism from $\emptyset$ to any other cover.
Let us for coverings $p_i : Y_i \to X$ define the finite coproduct to be the disjoint union $\bigsqcup_{i \in I} Y_i$ that inherits the topology of all the spaces. This is a cover, as all the p_i are covering maps and since $p\big|_{X_i} = p_i$ is continuous on any X_i and the X_i are disjoints, p is continuous on all of X.

Now, for some covering space Y of X let $G \subset Aut(Y)$ be a finite subgroup. Let us then define the quotient $Y/G = \{\{g(y) : g \in G\} : y \in Y\}$ together with the quotient topology and quotient map q. The map $f : Y/G \to X$ is surjective, as the covering map $p : Y \to X$ and quotient map q are surjective. It is continuous as Y/G is equipped with the quotient topology and therefore f is a restriction. Lastly, it is a covering, because for any open $U \subset X$:

$$\begin{aligned} & \bigcup_{\alpha \in I} V_\alpha = q^{-1}(U) = p^{-1}(f^{-1}(U) \\ \Rightarrow \quad & p\left(\bigcup_{\alpha \in I} V_\alpha\right) = \bigcup_{\alpha \in I} (p(V_\alpha)) = f^{-1}(U) \end{aligned}$$

and since p is a quotient map, we split up $f^{-1}(U)$ into open sets $p(V_\alpha)$. The homeomorphic attribute is trivial as q is a covering map.

G3. Let $u : Y_1 \to Y_2$ be a morphism. Let $u'' : Y_1 \to \{u^{-1}(y_2) : y_2 \in Im(u) \subset Y_2\}$ be a morphism such that $y_1 \mapsto f^{-1}(y_2)$ for that unique y_1 with $y_1 \in f^{-1}(y_2)$. This map is obviously surjective and therefore an epimorphism. Then the map $u' : \{u^{-1}(y_2) : y_2 \in u(Y_2)\} \to Y_2$, with $S \mapsto y_2$ if $u(a) = y_2$ for all $a \in S$. This is injective, since we factor out all the multiple mappings. As a result, it is a monomorphism, so we found u' and u'' that satisfy $u = u' \circ u''$.
What remains to be checked is that $\{u^{-1}(y_2) : y_2 \in u(Y_2)\}$ together with the map $q = p_2 \circ u'$ is indeed a covering of X, as the fact that u' and u'' are morphisms is obvious.
Let $U \subset X$ be a sufficiently small open set from definition 1.5.1. Then it holds that

$$q^{-1}(U) = u'^{-1} \circ p_2^{-1}(U) \overset{1.5.1}{=} u'^{-1}\left(\bigcup_{\alpha \in I} V_\alpha\right) = \bigcup_{\alpha \in I} (u'^{-1}(V_\alpha))$$

and since u' is continuous we know that $u'^{-1}(V_\alpha)$ is open for each α. This means that we can split up $q^{-1}(U)$ into disjoint open sets that q maps homeomorphically onto U, since $q = p_2 \circ u'$ and p_2 is a covering map.
Assume, $f : Y_1 \to Y_2$ is a monomorphism, meaning that f is injective, continuous and satisfies $p_1 = p_2 \circ f$. Let us then define $W = Y_2 \setminus f(Y_1)$ with the subspace topology. Then $W \sqcup f(Y_1) = Y_2$, since $f(Y_1) \subset Y_2$, which by definition 1.1.15 makes f a direct summand.

G4. Taking into account $F_x(id : X \to X) = id^{-1}(\{x\}) = \{x\}$ and our knowledge from chapter 2.1, the singletons are the final objects in the category *sets*; thus F_x maps final objects to final objects.
For the fiber product, let (Y_1, p_1) and (Y_2, p_2) be two coverings of X. Let $f : Y_1 \to Z$ and $g : Y_2 \to Z$ be two morphisms. Then, since $h = p_1 \times p_2$ is a covering map:

$$\begin{aligned} F_x(h : Y_1 \times_Z Y_2) = h^{-1}(\{x\}) &= \{(y_1, y_2) : f(y_1) = g(y_2) \ \wedge \ h((y_1, y_2)) = x\} \\ &= \{(y_1, y_2) : f(y_1) = g(y_2)\} \ \cap \ p_1^{-1}(\{x\}) \times p_2^{-1}(\{x\}) \\ &= p_1^{-1}(\{x\}) \ \times_{F_x(Z)} \ p_2^{-1}(\{x\}) \\ &= F_x\left(p_1 : Y_1 \to X\right) \ \times_{F_x(Z)} \ F_x\left(p_2 : Y_2 \to X\right) \end{aligned}$$

which implies that the functor and the fiber product commute.

G5. We now show that it commutes with finite coproducts. Let (X_i, p_i), $i \in [n]$ be coverings and I a finite set. Then

$$\begin{aligned} F_x(p : \bigsqcup_{i \in [n]} X_i \to X) = p^{-1}(\{x\}) &\overset{X_i \text{ disjoint}}{=} p_1^{-1}(\{x\}) \sqcup \cdots \sqcup p_n^{-1}(\{x\}) \\ &= F_x(p_1 : X_1 \to X) \sqcup \cdots \sqcup F_x(p_n : X_n \to X). \end{aligned}$$

Now, let us show that F_x maps epimorphisms to epimorphisms. For this matter, let us define $f : Y_1 \to Y_2$ to be an epimorphism, i.e. a surjective map such that $p_2 \circ f = p_1$. In that case, $F_x(f : p_1 \to p_2) = \overline{f} : p_1^{-1}(\{x\}) \to p_2^{-1}(\{x\})$, where $\overline{f}$ denotes the image of f under F_x, clearly gives us a surjective map, since f is surjective and p_1 and p_2 are coverings. Eventually, let us prove that F_x commutes with quotients. Let Y/G be a quotient and let $p_G : Y/G \to X$ be its covering map, where $p : Y \to X$ denotes the regular covering. Then:

$$\begin{aligned} F_x(p_G : Y/G \to X) &= p_G^{-1}(\{x\}) = \{G(y) : p_G(G(y)) = x\} \\ &= \{G(y) : p(y) = x\} = p^{-1}(\{x\})/F_x(G) = F_x(p : Y \to X)/F_x(G) \end{aligned}$$

where $F_x(G)$ is in some sense a restriction of an element in G.

G6. Let $u : Y_1 \to Y_2$ be a morphism and thus a continuous map with $p_1 = p_2 \circ u$. Then $\overline{u}$, the image of u under F_x, is the restriction of u to the fibers $p_1^{-1}(\{x\})$ and $p_2^{-1}(\{x\})$ and thus we get the map $\overline{u} : p_1^{-1}(\{x\}) \to p_2^{-1}(\{x\})$. Let us assume that $\overline{u}$ is an isomorphism, meaning that it is bijective. We want to show that u is bijective. From the definition of a covering, we see that each of the sets $\{y \in X : F_y(u) \text{ bijective}\}$ and $\{y \in X : F_y(u) \text{ not bijective}\}$ is open. Since X is connected, one of them is X and the other one is $\emptyset$, but since we assumed that for one $x \in X$ the image of u under F_x is bijective, we conclude that $F_y(u)$ is always bijective. As u is bijective on all of its fibers, it is bijective everywhere, which implies that u is an isomorphism. □

Having proved this, let us proceed with various non-categorial notions about covering spaces.

3.1 Universal Cover

In this section we introduce a similar notion to the *absolute Galois group* that we have dealt with in the second chapter, only this time for covering spaces. This notion will be called an *universal cover.* Before being able to do so, let us recall some basic notions from topology. A topological space X is *connected*, if there are no two disjoint open sets U and V such that $U \cup V = X$. It is called *locally path-connected*, if for every $x \in X$ and every open neighborhood U of x there is an open neighborhood $V \subset U$ of x that is *path-connected*, i.e. where for any two points $x_1, x_2 \in V$ there is a path connecting them. Finally, X is called *semi-locally simply connected*, if every point X has an open neighborhood U, such that any loop in U is homotopic to a constant path e_x in some point x in X. When we compare this to definition 1.5.2, we see that it is indeed different from U being simply connected, because the contraction of the loop does not have to take place in U. This definition seems quite unusual but we will later on observe an example that is semi-locally simply-connected, yet not locally simply-connected.
This eventually gives us all the means to formulate the following definitions:

Definition 3.1.1 For a topological space X a covering space $(\tilde{X}, q)$ is called a *universal cover*, if it is simply-connected.

Example: The universal cover of the punctured complex plane $\mathbb{C}^\star$ is the complex plane itself. Here, the covering map is given my the exponential function $\exp : \mathbb{C} \to \mathbb{C}^\star$. Due to the fact that exp is holomorphic on the entire complex plane, we can find a neighborhood U around any point x of $\mathbb{C}^\star$ where the diameter of the preimage of U is stricly smaller than 2π. If necessary, we can always shrink the open set U to ensure this requirement. Since exp is $2\pi i$-periodic, the preimage splits into an infinite amount of copies V_k that differ only in shiftings of multiples of $2\pi i$. By the basic fact that exp is a biholomorphism as a restriction to $(-\pi, \pi)$, or any other open interval with length 2π, it is a biholomorphism on each V_k which proves that it is a covering map.

The fact that $\mathbb{C}$ is simply-connected is trivial. The deck transformation group $Aut(\mathbb{C}/\mathbb{C}^\star)$ contains all the maps $z \mapsto z + 2\pi i$ and is thus isomorphic to $\mathbb{Z}$, exactly as the fundamental group $\pi_1(\mathbb{C}^\star)$.

Definition 3.1.2 Let $p_1 : C_1 \to X$ and $p_2 : C_2 \to X$ be two covers of a topological space X. If there is a homeomorphism $f : C_1 \to C_2$ such that $p_2 \circ f = p_1$, then C_1 and C_2 are called *equivalent.*

Proposition 3.1.3 *If the underlying topological space X is connected, locally path-connected and semi-locally simply connected, then X has a universal cover.*

PROOF For the proof see Koch, [8], pages 1f.

Example: As most of the path-connected and locally path-connected topological spaces that we encounter in our daily life are actually semi-locally simply-connected, let us now view an example that is locally path-connected, path-connected, but not semi-locally simply-connected. This is the *Hawaiian Earring Group* $\mathcal{H}$ that is defined by

$$\mathcal{H} = \bigcup_{n \in \mathbb{N}} C_n$$
$$\text{where } C_n = \{(x, y) \in \mathbb{R}^2 : (x - 1/n)^2 + y^2 = 1/n^2\}.$$

Since all circles share the common point $(0, 0)$ and each C_n itself is path-connected, $\mathcal{H}$ is path-connected.

If we pick any point x in $\mathcal{H} \setminus (0, 0)$, let U_x be any neighborhood of x. We know that x lies on some C_n. Thus we can find a sufficiently small ball around x that does not intersect with any other C_m with $m \neq n$ and is contained in U_x. Since every C_k is path-connected, $\mathcal{H} \setminus (0, 0)$ is locally path-connected. Inversely, from 0, taking any neighborhood U_0 in $\mathcal{H}$ around 0 and any point y in U_0, we know that y needs to lie on some C_n. Since every C_n is path-connected, there exists a path from 0 to y and therefore, U_0 is path-connected. This implies that $\mathcal{H}$ is locally-path-connected. The figure 3.1 depicts $\mathcal{H}$.

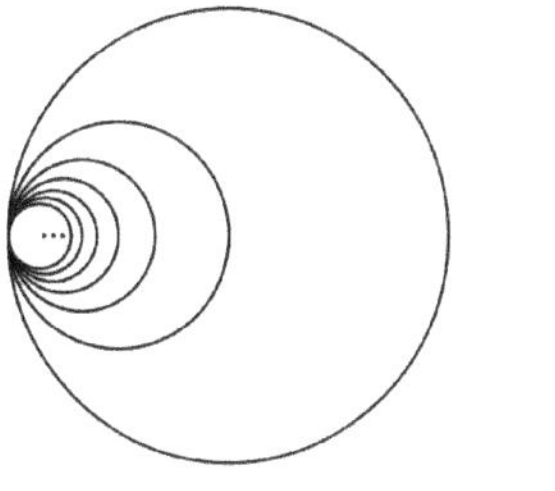

Figure 3.1: Hawaiian Earring Group [1]

Since every neighborhood of 0 contains a C_n and C_n is not simply-connected, we know that $\mathcal{H}$ is not semi-locally simply-connected. Obviously, the fundamental group of $\mathcal{H}$ is going to be quite "wild", but we are not going into detail here.

In fact, $\mathcal{H}$ has no universal cover. This we can see by assuming it had a universal cover $\tilde{\mathcal{H}}$ with covering map $p : \tilde{\mathcal{H}} \to \mathcal{H}$. We then view 0 and its lift to $\tilde{\mathcal{H}}$ that we denote by x_0. By definition 1.5.1 there is a neighborhood U of 0 that is evenly covered, so that $p^{-1}(U)$ splits into open sets $\{V_\alpha\}$ that each V_α is mapped homeomorphically into U by p.

Let $i : U \to X$ be the inclusion map. Then for the fundamental groups, since p maps makes U and V_α topologically equivalent, it holds that $\pi_1(V_\alpha) \cong \pi_1(U) \xrightarrow{i_\star} \pi_1(\mathcal{H})$ and the inclusion $j : V_\alpha \to \tilde{\mathcal{H}}$ gives us $\pi_1(V_\alpha) \xrightarrow{j_\star} \pi_1(\tilde{\mathcal{H}}) \xrightarrow{p_\star} \pi_1(\mathcal{H})$. These two compositions are equivalent. As $\tilde{\mathcal{H}}$ is simply-connected by definition 3.1.1, $p_\star$ is the zero-map and therefore, $i_\star$ is the zero-map. We

[1] https://wildtopology.wordpress.com/2013/11/23/the-hawaiian-earring/

can conclude that U is simply-connected that makes $\mathcal{H}$ semi-locally simply-connected which is a contradiction to our assumption. Thus, $\tilde{\mathcal{H}}$ has no universal cover.

Theorem 3.1.4 *Let X be a connected, locally path-connected and semi-locally simply-connected topological space with universal cover $(\tilde{X}, q)$. Fix points $x \in X$ and $d \in \tilde{X}$ such that $q(d) = x$. Let $r : E \to X$ be another path-connected cover and $e \in E$ a point with $r(e) = x$. Then there exists a unique covering map $f : \tilde{X} \to E$ such that $r \circ f = q$ and $f(d) = e$. In other words, the universal cover covers any other cover.*

PROOF First off, by proposition 3.1.3 the universal cover of X exists and let us call it $p : \tilde{X} \to X$. Let $r : E \to X$ be some other connected cover. By proposition 1.5.3 and the fact that the cover is path-connected, for the fundamental group of $\tilde{X}$ it holds that $\pi_1(\tilde{X}) = \{0\}$. Then $p_\star(\pi_1(\tilde{X})) = \{0\}$, because $p_\star$ is a homomorphism. Hence, the condition of the Lifting criterion 1.5.6 is fulfilled and we can lift $p : \tilde{X} \to X$ to a unique continuous map $f : \tilde{X} \to E$, such that $r \circ f = p$ and $f(d) = e$.

We still need to show that f is surjective and fulfills definition 1.5.1 for being a covering map. For that, we introduce the below commutative diagram that will help us understand the following.

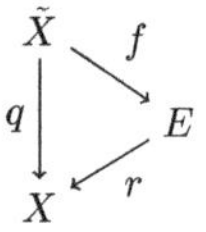

Let $\tilde{x}_0 \in \tilde{X}$. We set $e_0 = f(\tilde{x}_0)$ and $x_0 = q(x_0)$ and pick a path $\tilde{\alpha} : e_0 \rightsquigarrow e$ in E for some $e \in E$. Then $\alpha = r \circ \tilde{\alpha}$ is a path in X beginning at x_0 and $\tilde{\tilde{\alpha}}$ a lifting of $\tilde{\alpha}$ that begins at $\tilde{x}_0$. By proposition 1.5.6 we get that $f \circ \tilde{\tilde{\alpha}}$ is a lifting to E and due to uniqueness this also yields that $\tilde{\alpha} = f \circ \tilde{\tilde{\alpha}}$. Moreover, f maps the end point of $\tilde{\tilde{\alpha}}$ to e which shows that for any choice of $e \in E$ we can find $\tilde{x} \in \tilde{X}$ such that $f(\tilde{x}) = e$, implying surjectivity.
Let us show that f is also a covering map. Given $e \in E$ we set $x = r(e)$. Because q and r are both covering maps and X is path-connected, there is a path-connected open neighborhood U_x of x for which it holds that

$$q^{-1}(U_x) = \bigcup_\alpha V_\alpha \subset \tilde{X} \quad \text{and} \quad r^{-1}(U_x) = \bigcup_\beta W_\beta \subset E$$
$$\text{with} \quad q : V_\alpha \xrightarrow{\sim} U_x \quad \text{and} \quad r : W_\beta \xrightarrow{\sim} U_x$$

homeomorphically. Then pick $W \in \{W_\beta\}$ with $e \in W$. Since q is a restricted homeomorphism and U_x is connected, because X is locally path-connected by assumption, every single V_α is connected and thus the image $f(V_\alpha)$ is also connected, since f is continuous. As each slice V_α is mapped into $r^{-1}(U_x)$ and they are connected, they are mapped into a single slice W_β. Now, we pick all the V_α that f maps into our fixed W from above and we denote this set by $\{V_i\}$. Then $f^{-1}(W) = \bigcup_i V_i$. It remains to be shown that those V_i are mapped homeomorphically into W, but this is clear: the restrictions $q\big|_{V_i} : V_i \to U_x$ and $r\big|_W : W \to U_x$ are both homeomorphisms by assumption and so the composition $f\big|_{V_i} = r\big|_W^{-1} \circ q\big|_{V_i}$ is also a homeomorphism. Eventually, this proves the theorem because f is a covering map. □

Corollary *The universal cover is unique up to equivalence of covering spaces, if it exists. Picking one representative of the equivalence class enables us to speak of the universal cover of X and we denote it by $\tilde{X}$.*

PROOF Let $q_1 : \tilde{X}_1 \to X$ and $q_2 : \tilde{X}_2 \to X$ be universal covers. Fix two arbitrary points $c \in \tilde{X}_1$ and $d \in \tilde{X}_2$ that are both mapped to some $x \in X$ by q_1 or q_2 respectively. By theorem 3.1.4 they cover each other, since they both are universal covers of X, so we get the following commutative diagram:

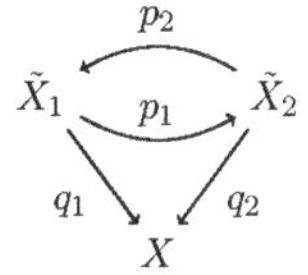

Here, p_1 also satisfies $p_1(c) = d$ and p_2 satisfies $p_2(d) = c$ by theorem 3.1.4. Since by the above diagram we have $q_1 \circ p_2 \circ p_1 = q_2 \circ p_1 = q_1$ and $q_2 \circ p_1 \circ p_2 = q_1 \circ p_2 = q_2$, this implies that the following diagrams

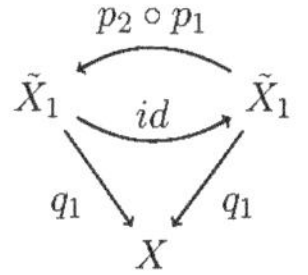

and

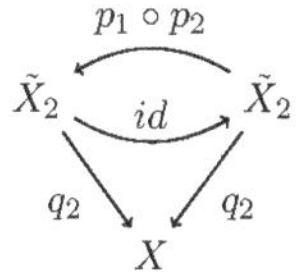

commute. Furthermore, $p_2 \circ p_1$ is a lifting of q_1 with $p_2 \circ p_1(c) = c$. This means by lemma 1.5.6 that it is unique. Since $id_{\tilde{X}_1}$ is another lifting of q_1 such that $id_{\tilde{X}_1}(c) = c$, the uniqueness of a lifting gives us that $p_2 \circ p_1 = id_{\tilde{X}_1}$.
A similar argument yields that $p_1 \circ p_2 = id_{\tilde{X}_2}$. Since p_1 and p_2 are continuous, this implies that $p_1 \circ p_2$ is a homeomorphism that is compatible with q_2. Thus, by definition 3.1.2, we know that $\tilde{X}_1$ and $\tilde{X}_2$ are equivalent. □

Example: Let us now, at the end of this chapter, show that the condition "semi-locally simply-connected" that seems quite arbitrary, is actually necessary and cannot be replaced by the condition "locally simply-connected", which means that around each point and each neighborhood of this point we can find a simply-connected neighborhood. The example here will be the cone that arises from the Hawaiian Earring Group $\mathcal{H}$ that we viewed earlier. This surface is the image of

$$c : \mathcal{H} \times [0,1] \to \mathbb{R}^3, \qquad c(x,y,t) = ((1-t)x, (1-t)y, t)$$

and roughly like it is depicted in figure 3.2:

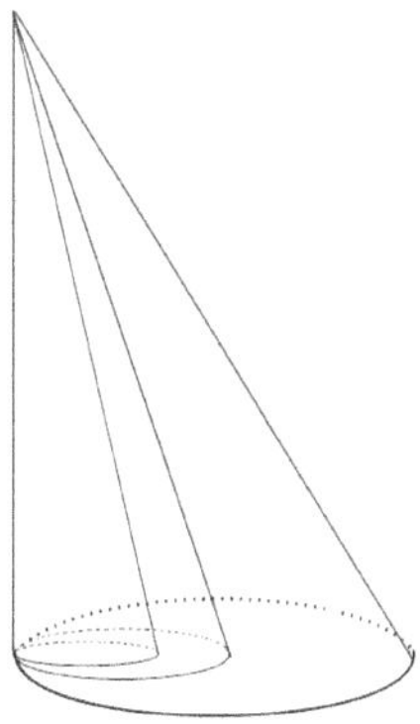

Figure 3.2: The cone of the Hawaiian Earring Group $\mathcal{H}$

This cone is obviously simply-connected and hence semi-locally simply-connected, since we can contract each path on it to the critical point $(0,0,1)$. However, choosing a sufficiently small neighborhood around the point $(0,0,0)$, there is always an entire circle of $\mathcal{H}$ contained in this neighborhood. Thus, this neighborhood is not simply-connected. This eventually implies that the cone over the Hawaiian Earring is not locally simply-connected.

3.2 Coverings with marked points

Definition 3.2.1 Let $p : Y \to X$ be a cover such that $p(y_0) = x_0$ for some points. We then call it a *covering with marked points* and denote it by $p : (Y, y_0) \to (X, x_0)$. For doing so, let us introduce a new definition. Two coverings with marked points $p_1 : (Y, y_0) \to (X, x_0)$ and $p_2 : (Z, z_0) \to (X, x_0)$ are called *equivalent*, if there exists a homeomorphism $f : Y \to Z$ such that $p_1 = p_2 \circ f$ and $f(y_0) = z_0$.

A *covering with marked points* $p : (Y, y_0) \to (X, x_0)$ defines the induced homomorphism $p_\star : \pi_1(Y, y_0) \to \pi_1(X, x_0)$ from definition 1.4.6. The following proposition further specifies this corellation.

Proposition 3.2.2 *If $p : (Y, y_0) \to (X, x_0)$ is a covering with marked points, then the induced group homomorphism $p_\star : \pi_1(Y, y_0) \to \pi_1(X, x_0)$ is injective.*

PROOF We show that the kernel of $p_\star$ is trivial. For that matter, let e_{x_0} be the constant path in $\pi_1(X, x_0)$. Let γ be a loop in $\pi_1(Y, y_0)$, whose equivalence class is mapped to e_{x_0}. We show that it is homotopic to e_{y_0}. Let us use the corollary of lemma 1.5.5. We define a homotopy H between $p \circ \gamma$ and e_{x_0}. If $\tilde{H}$ is a lifting of H to Y such that $\tilde{H}(0,0) = y_0$, then $\tilde{H}$ is a path homotopy between γ and the constant loop e_{y_0}. □

This notion implies that, since the image of a group under a group-homomorphism is a subgroup of the codomain, every covering with marked points correlates with a subgroup of the fundamental group $\pi_1(X, x_0)$. The following theorem gives rise to this:

Theorem 3.2.3 (On the classification of coverings with marked points, [6]) *For every subgroup G of the fundamental group $\pi_1(X, x_0)$ of a space X there exists a connected covering with marked points $p : (Y, y_0) \to (X, x_0)$ such that $p_\star(\pi_1(Y, y_0)) = G$.*

PROOF For the proof of this theorem I refer to Khovanskii, [6], page 44.

Lemma 3.2.4 *Path-connected coverings with marked points of a space X are equivalent as coverings, if and only if the subgroups corresponding to these coverings are conjugate in $\pi_1(X, x_0)$.*

PROOF Let $p_1 : (Y, y_0) \to (X, x_0)$ and $p_2 : (Z, z_0) \to (X, x_0)$ be equivalent as coverings. This means that there exists a homeomorphism $h : Y \to Z$ such that $p_1 = p_2 \circ h$. This implies that h necessarily maps the fiber $p_1^{-1}(x_0)$ to $p_2^{-1}(x_0)$ which, in turn, implies that $p_1 : (Y, y_0) \to (X, x_0)$ is equivalent to $p_2 : (Z, h(y_0)) \to (X, x_0)$ as a covering with marked points. This also gives us that $p_2(h(y_0)) = p_2(z_0) = x_0$. Now, let $\tilde{\gamma}$ be a path in Y that starts at z_0 and ends at $h(y_0)$. This is possible, since Y is path-connected by assumption. For this path, it holds that $p \circ \tilde{\gamma} = \gamma$ is a loop that starts at x_0 and ends at x_0.

Let us define G_1 to be the subgroup of $\pi_1(X, x_0)$ that consists of the paths, whose lifts to Z start and end at $h(y_0)$.

Now, given a path $\delta \in p_{2\star}(\pi_1(Z, z_0))$, we know that the lift of $\gamma \star \delta \star \overline{\gamma}$ starts and ends at $h(y_0)$ and thus it is in G_1. For the other inclusion, we have that $\overline{\gamma} \star \eta \star \gamma$ is in $p_{2\star}(Z, z_0)$ for some $\eta \in G_1$, since its lift starts and ends at z_0. Therefore, if $G = p_{2\star}(\pi_1(Z, z_0))$, we know that $\gamma \star G \star \overline{\gamma} = G_1$ which means exactly that G and G_1 are conjugate. This completes the first implication, since h is a homeomorphism between covers with marked points and therefore $p_{1\star}(\pi_1(Y,y_0)) = p_{2\star}(\pi_1(Z,h(y_0)))$.

For the second implication, if the corresponding subgroups are conjugate in $\pi_1(X,x_0)$, there exists a homeomorphism with marked points between (Z,z_0) and (Y,a_0) for some point $a_0 \in p_1^{-1}(x_0)$ that is potentially different from y_0. But with the argument from above, (Y,a_0) and (Y,y_0) are equivalent as covering spaces. □

This lemma gives us the following interpretation: Connected covers $p: Y \to X$ of a connected, locally path-connected and semi-locally simply connected space X are classified by the subgroups of the fundamental group $\pi_1(X,x_0)$ that are defined up to conjugation in the group $\pi_1(X)$. Now, in Galois theory, we normally view algebraic field extensions that are subordinate to a given Galois extension and compare them to the subgroups of the Galois group. For that matter, compare 2.2.6. To do something similar here, let us introduce new notions.

Definition 3.2.5 We say that a covering with marked points $p_2 : (Y_2, y_2) \to (X, x_0)$ is *subordinate* to a cover $p_1 : (Y_1, y_1) \to (X, x_0)$, if there is a morphism $h : (Y_1, y_1) \to (Y_2, y_2)$ such that $p_1 = p_2 \circ h$. We call a connected cover $p : (Y, y_0) \to (X, x_0)$ *normal*, if it corresponds to a normal subgroup of the fundamental group $\pi_1(X, x_0)$.

Lemma 3.2.6 *For a normal subgroup H of $\pi_1(X,x_0)$ the group of deck transformations of the corresponding normal covering $p: Y \to X$ is isomorphic to the group $\pi_1(X,x_0)/H$.*

PROOF Following Khovanskii, [6], we know that if a deck transformation takes y_0 to y_0', then it is unique, because the set on which two deck transformations coincide is open, since p is a local homeomorphism as a covering map, it is closed, since its complement $\{y : \sigma_1(y) \neq \sigma_2(y)\}$ is open by the definition of a covering map and the fact that deck transformations permute fibers. Finally, it is non-empty, because we know that y_0 is in this set. Since Y is connected that means that those deck transformations coincide on all of Y.
Now, the fundamental group $\pi_1(X,x_0)$ acts by product from the right (compare definitions 1.4.2 and 1.2.6) on the set $\Omega(X,x_0)$ of all paths starting at x_0 modulo the equivalence relation, where two paths are equivalent if and only if they end at the same point. This makes us view the space of orbits under the action of the normal subgroup $H \subset \pi_1(X,x_0)$ on $\Omega(X,x_0)$, and denote it by $\Omega_H(X,x_0) = \{\gamma \star H : \gamma \in \Omega(X,x_0)\}$ and also multiply it from the right with $\pi_1(X,x_0)$. Now, the equivalence class $x \star H$ is sent to $x \star H \star \gamma$ via multiplication with $\gamma \in \pi_1(X,x_0)$. This is equal to $x \star \gamma \star H$, since H is normal. Using the projection $f : \Omega_H(X,x_0) \to (X,x_0)$ that assigns to each path the point it terminates, we see that it is compatible with the action of the fundamental group on $\Omega_H(X,x_0)$, so the fundamental group acts on the space Y via deck transformations. Assuming Y is the covering that corresponds to the normal subgroup H of $\pi_1(X,x_0)$, this gives us that H is the kernel of this action and therefore $Aut(Y/X) \cong \pi_1(X,x_0)/H$. □

Theorem 3.2.7 *Let $p : (Y,y_0) \to (X,x_0)$ be a normal covering. There is a bijection*

$$\{(M,m_0) : (M,m_0) \text{ is a subordinate connected covering}\} \overset{\sim}{\longleftrightarrow} \{H : H \text{ subgroup of } Aut(Y/X)\}$$

Furthermore, a subordinate covering (M,m_0) of X is normal if and only if it corresponds to a normal subgroup H of $Aut(Y/X)$. It then holds that $Aut(M/X) = Aut(Y/X)/H$.

PROOF Let $N = \pi_1(X,x_0)/H$ be the deck transformation group that corresponds to the normal cover induced by the normal subgroup H. Let $f : (M,m_0) \to (X,x_0)$ be a subordinate covering and G the corresponding image $f_\star(\pi_1(M,m_0))$ in $\pi_1(X,x_0)$. With this, we associate the subgroup of the deck transformation group N that is equal to the image of G under the quotient group homomorphism $\pi_1(X,x_0) \to N = \pi_1(X,x_0)/H$.
Since for a chain of subordinate covers $Y \overset{q}{\to} \dots \overset{p_{n+1}}{\to} M_n \overset{p_n}{\to} \dots \overset{p_2}{\to} M_1 \overset{p_1}{\to} X$ it holds that

$$p_{1\star} \circ \dots \circ q_\star(\pi_1(Y,y_0)) \subset \dots \subset p_{1\star} \circ \dots \circ p_{n\star}(\pi_1(M,m_0)) \subset \dots \subset \pi_1(X,x_0)$$

we have that this map is bijective.

For the second part, a subordinate cover is normal by definition 3.2.5, if it corresponds to a normal subgroup L of $\pi_1(X, x_0)$. Then, in turn, by the above construction, $\pi_1(X, x_0)/L$ is a subgroup of $Aut(Y/X) = \pi_1(X, x_0)/H$. It is normal, since H is normal in L. With the group homomorphism $Aut(Y/X) \to Aut(M/X)$, $\sigma \mapsto p \circ \sigma$, where $p : Y \to M$ is the morphism from definition 3.2.5, we see that the H from the construction above is the kernel of this map and therefore we have that $Aut(M/X) \cong Aut(Y/X)/H$. □

Comparing this result to the fundamental theorem of Galois theory by Krull (theorem 2.2.6), gives the impression that the fundamental group and the Galois group are closely related, though one may not be able to understand why. In chapter 4, we will see a specific example and compare the two entities.

Lemma 3.2.8 *Let X be a path-connected, locally path-connected and furthermore a semi-locally simply-connected topological space. Then the universal covering $\tilde{X}$ is a normal covering of X.*

PROOF We note that $\{e_{x_0}\}$ is a normal subgroup of $\pi_1(X, x_0)$ and that $\tilde{X}$ corresponds to $\{e_{x_0}\}$, since it is simply-connected. Therefore, it is a normal covering of X. □

Corollary *Let X be a path-connected, locally path-connected and semi-locally simply-connected topological space. Then there is an inclusion-reversing bijective correspondence between subgroups of $\pi_1(X, x_0)$ and connected coverings of X.*

PROOF Since $\tilde{X}$ is simply-connected, its fundamental group corresponds to the trivial fundamental group in $\pi_1(X, x_0)$. By Lemma 3.2.8 we get that $q : \tilde{X} \to X$ is a normal covering. Those two properties of $\tilde{X}$ together enable us to apply theorem 3.2.7, so that there is a bijective correspondence between subgroups of the group of deck transformations $Aut(\tilde{X}/X)$ and the subordinate coverings of $\tilde{X}$. However, by the construction above, any deck transformation group is isomorphic to a quotient of $\pi_1(X, x_0)$ by a normal subgroup. Since the subgroups of $Aut(\tilde{X}/X)$ are induced by restrictions of its elements, there is a one-to-one correspondence between the subordinate coverings of $\tilde{X}$ and the subgroups of $\pi_1(X, x_0)$. □

3.3 The profinite completion of the Fundamental Group

Let $p_i : Y_i \to X$ be a normal finite covering of a path-connected, locally path-connected and semi-locally simply-connected space X. Let $\tilde{X}$ be its universal cover that exists by proposition 3.1.3. The partially ordered set is induced by $i \succ j \Leftrightarrow p_j : Y_j \to X$ is a subordinate covering of $p_i : Y_i \to X$, because then $Aut(Y_i/X) \supset Aut(Y_j/X)$. By definition 1.2.3, this is a projective system with projective limit $\hat{\pi}_1 = \varprojlim Aut(Y_i/X)$ which, by proposition 1.2.4 is a profinite group equipped with the subspace topology as a subset of $\prod_i Aut(Y_i/X)$.

By lemma 3.2.6, $Aut(Y_i/X) \cong \pi_1(X)/H$ for the normal subgroup H of $\pi_1(X)$ that corresponds to the normal finite covering $p_i : Y_i \to X$. Here, H has finite index, because p_i is a finite covering. This makes $\hat{\pi}_1$ the profinite completion of $\pi_1(X)$ and justifies the name.
Using a remark from chapter 1.2, there is a natural group homomorphism $\mu : \pi_1(X) \to \hat{\pi}_1$ that satisfies the following property: for each profinite group H and for any group homomorphism $f : \pi_1(X) \to H$ there is a continuous group homomorphism $g : \hat{\pi}_1 \to H$ such that $f = g \circ \mu$.

Proposition 3.3.1 *The image $im(\mu)$ of $\pi_1(X)$ is a dense subset of $\hat{\pi}_1$.*

PROOF Let $U \subset \hat{\pi}_1$ be a non-empty open set. By the product topology of sets with the discrete topology and the subspace topology we get that $U = \prod_N U_N \cap \hat{\pi}_1$, where U_N is equal to $\pi_1(X)/N$ for all but finitely many U_N. Without loss of generality, we assume that there exist proper subsets

U_N, otherwise there is nothing to prove.

Accordingly, let us denote $N_1, \ldots, N_m$ the normal subgroups of finite index, where we have $U_{N_i} \neq \pi_1(X)/N_i$. We define $M = \bigcap_i N_i$ which is also a normal subgroup of finite index. Let us now view the canonical group homomorphisms $\mu_{MN_i} : \pi_1(X)/M \twoheadrightarrow \pi_1(X)/N_i$ that are well defined, since $M \subset N_i$ and are induced by definition 1.2.3 of the projective system that we view here. Consequently, we define $V_i = \mu_{MN_i}^{-1}(U_{N_i}) \subset \pi_1(X)/M$.

By assumption, there exists $(x_N) \in \prod_N U_N \cap \hat{\pi}_1$ indicating that $V = \bigcap_i V_i$ is not empty because for the (x_N) from above we have that μ_{MN_i} satisfies $\mu_{MN_i}(x_M) = x_{N_i}$. Thus, $x_M \in V$, because $\mu_{MN_i}^{-1}(x_{N_i}) \subset V_i$ contains x_M for all i. Then, using the quotient map $p_M : \pi_1(X) \twoheadrightarrow \pi_1(X)/M$, if we pick $\gamma \in V$ such that $p(\gamma) \in V$, we get that $\mu(\gamma) \in \prod_N U_N \cap \hat{\pi}_1$, using $\mu : \pi_1(X) \to \hat{\pi}_1$ from the proposition. Therefore, $im(\mu)$ is a dense subset of $\hat{\pi}_1$. □

This, at least, puts $\hat{\pi}_1$ in a context with $\pi_1(X)$. We now know that the closure of $im(\mu)$ is $\hat{\pi}_1$, since it is a dense subset.

In the following, we will try to identify the profinite groups $\tilde{\pi}_1$ and $\tilde{\pi}_2$ that make the Galois categories $\mathcal{C}_k$ from proposition 2.3.5 and $\mathcal{C}_X$ from a claim in chapter 3 $\tilde{\pi}_1$-sets or respectively $\tilde{\pi}_2$-sets. This notion arises from proposition 2.1.3.

First, it is clear that $\mathcal{C}_X$ is an essentially small category, because the fundamental group π_1 is a set. Since all subordinate coverings of the universal cover correspond to a quotient of π_1, the collection of all quotients of π_1 by a normal subgroup also forms a set. As the objects of $\mathcal{C}_k$ correspond to normal subgroups of finite index, it is a subset of quotients of π_1 by a normal subgroup and thus a set itself. Consequently, both categories fulfill the requirements of proposition 2.1.3 and thus we can use it.

By a construction of Lenstra, [5], found on pages 35f., that was already used in chapter 2, we know that $\tilde{\pi}_1$ is isomorphic to $\lim\limits_{\leftarrow} Aut_{\mathcal{C}_X}(A)$, where A runs over all connected Galois objects in a suiting way. In this case, a connected Galois object corresponds to a connected normal covering, so using the construction from the beginning of this chapter, we get that $\tilde{\pi}_1 \cong \widehat{\pi_1(X)}$. This makes $\mathcal{C}_X$ equivalent to the category finite $\tilde{\pi}_1$-*sets*.

Moreover, by the second chapter, $Gal(k)$ by theorem 2.3.6 acts on $Hom_{{}_kSAlg}(A, \overline{k})$ for a finite étale algebras A. This way, we get that the category $\mathcal{C}_k$ is equivalent to $Gal(k)$-sets, since ${}_kSAlg$ is anti-equivalent to $Gal(k)$-sets.

This makes us compare those two entities and understand that $\mathcal{C}_X$ and $\mathcal{C}_k$ are of the same form. The only difference here is the profinite group.

Chapter 4

Riemann Surfaces

Let us now take a look at a more concrete example that brings the mentioned concepts together, more specifically complex manifolds. First, we are going to give definitions and prove basic propositions. Eventually, this will yield a theorem that not only compares the fundamental group and the Galois group, but it will even give a correspondence between them, specifically on Riemann Surfaces.

4.1 Riemann Surfaces

Definition 4.1.1 Let X be a Hausdorff topological space. A *holomorphic atlas* $\mathcal{A} = \{(U_i, h_i)\}$ is made up by *charts* (U_i, h_i), consisting of open sets $U_i \subset X$ that cover X and homeomorphisms $h_i : U_i \to h_i(U_i) \subset \mathbb{C}$ that are holomorphically compatible in the following sense: for every i the set $h_i(U_i)$ is open in $\mathbb{C}$ and for every i, j it holds that the *transition maps*

$$h_j \circ h_i^{-1} : h_i(U_i \cap U_j) \to h_j(U_i \cap U_j)$$

are biholomorphic in the classic sense.

This property is displayed in figure 4.1:

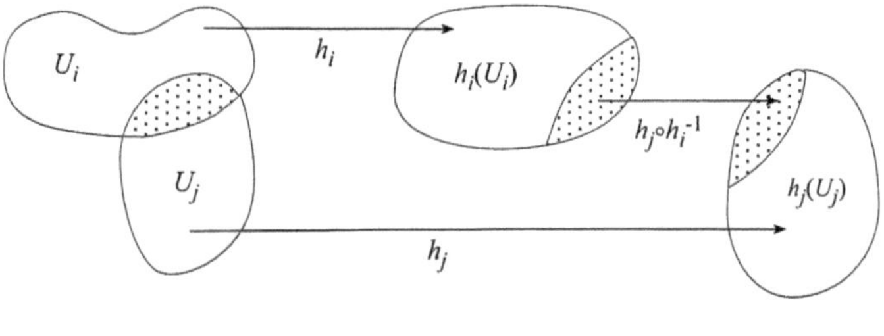

Figure 4.1: The Transition Maps $h_j \circ h_i^{-1}$ are biholomorphic [1]

Definition 4.1.2 Two atlases $\mathcal{A}$ and $\mathcal{B}$ are called *equivalent*, if $\mathcal{A} \cup \mathcal{B}$ is a holomorphic atlas. An equivalence class of atlases is called *holomorphic structure*. A *Riemann Surface* is a Hausdorff space X together with a holomorphic structure on X.
Furthermore, a *compact Riemann surface* is a Riemann surface X, such that any open cover of X has a finite subcover.

[1]Klaus Lamotke, Riemannsche Flächen, page 3.

Let us now treat an example to understand how to actually show that a topological space is a Riemann surface.

Example: Consider the complex projective line $\mathbb{C}P^1 = \mathbb{C} \cup \{\infty\}$ that is also called Riemann sphere and is obtained by identifying non-zero elements in $\mathbb{C}^2$ that are multiples of each other. Let us pick the charts $h_1 : z \mapsto z$ on $\mathbb{C}P^1 \setminus \{\infty\}$ and $h_2 : z \mapsto z^{-1}$ on $\mathbb{C}P^1 \setminus \{0\}$. They are indeed holomorphically compatible, since the transition maps $h_1 \circ h_2^{-1}(z) = 1/z = h_2 \circ h_1^{-1}(z)$ are holomorphic on $\mathbb{C}P^1 \setminus \{0, \infty\}$. This makes it a compact Riemann surface, as it is the one-point compactification of the Hausdorff topological space $\mathbb{C}$. It is homeomorphic to the sphere S^2,

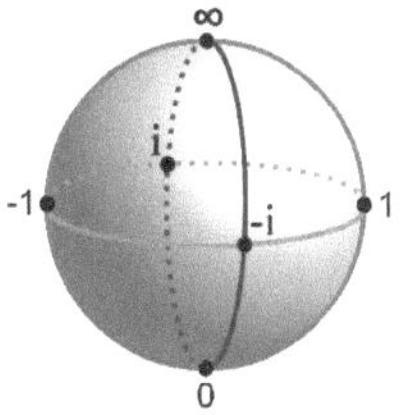

Figure 4.2: Depiction of the Riemann Sphere [2]

since the stereographic projection is a homeomorphism $\pi : S^2 \to \mathbb{C}P^1$. This correlation is depicted in figure 4.2 and it implies that $\mathbb{C}P^1$ is both compact and connected, as S^2 has these properties.

Definition 4.1.3 Let X be a Riemann surface and $U \subset X$ open. The map $f : U \to \mathbb{C}$ is called *holomorphic with respect to* $\mathcal{A}$, if for every chart $(V_i, h_i) \in \mathcal{A}$ the function

$$h_j(U \cap V_j) \to \mathbb{C}, \quad z \mapsto f \circ h_j^{-1}(z)$$

is holomorphic in the classic sense.

Remark: All the maps that are holomorphic with respect to $\mathcal{A}$, form a ring $\mathcal{O}(U, \mathcal{A})$. All the functions in $\mathcal{O}(U, \mathcal{A})$ are continuous. If we fix a holomorphic structure on X, we get for any open set $U \subset X$ the $\mathbb{C}$-algebra $\mathcal{O}(U) = \mathcal{O}(U, \mathcal{A})$, as on equivalence classes this ring does not depend on the choice of an atlas.

Definition 4.1.4 Let X and Y be Riemann Surfaces and $\mu : X \to Y$ a continuous function. μ is called *holomorphic*, if for any open set $U \subset Y$ and any function $f \in \mathcal{O}(U)$ it holds that $f \circ \mu \in \mathcal{O}\left(\mu^{-1}(U)\right)$.

Theorem 4.1.5 (Riemann's Existence Theorem, [1])
Let X be a compact Riemann surface, $x_1, \ldots, x_n \in X$ and $a_1, \ldots, a_n \in \mathbb{C}$. Then we can find $f \in \mathcal{M}(X)$ such that f is holomorphic at all the x_i and $f(x_i) = a_i$.

PROOF This theorem was taken from Szamuely, [1], page 72, and will not be proved here.

Proposition 4.1.6 *If the space X is a Riemann surface, every connected covering $p : D \to X$ inherits a canonical structure of a Riemann surface and $p : D \to X$ is a holomorphic map.*

PROOF Let X be a Riemann Surface and $y \in D$ arbitrary but fixed. Let (U, ϕ) be a chart around $p(y)$ that exists due to 4.1.1. If necessary, we can take smaller neighborhoods U of $p(y)$, so we are able to assume that by p being a covering map there exists an open set $V \subset D$ that is mapped homeomorphically to U. Then $(V, \phi \circ p)$ defines a chart around y in D. By taking all the open

[2] `https://en.wikipedia.org/wiki/Riemann_sphere`

sets V that cover D, these charts define an atlas due to the fact that $p : V_i \cap V_j \mapsto U_i \cap U_j$ homeomorphically.
For the transition maps, this gives us $(\phi_i \circ p) \circ (\phi_j \circ p)^{-1} = (\phi_i \circ p) \circ (p^{-1} \circ \phi_j^{-1}) = \phi_i \circ \phi_j$. By X being a Riemann surface, this is biholomorphic. This makes D a Riemann surface and, since by the above equation p is locally represented by the identity, p is holomorphic, as the identity is a holomorphism. □

4.2 Meromorphic Functions

Definition 4.2.1 $S \subset X$ is called a set of isolated points, if for each $x \in S$ there exists an open neighborhood $U \subset X$ of x, such that $U \cap (S \setminus \{x\}) = \emptyset$.

Remark: If we assume X to be compact relative to the topological space it lies in, then any set of isolated points is finite. The definition of a set of isolated points gives us a natural covering by filling X with more open sets that do not intersect with S (X is Hausdorff). Since S is already covered, we can use the compactness to achieve a finite subcovering. Nevertheless, S was only covered by the opens from definition 4.2.1, so they all have to appear in this subcovering and thus, there are only finitely many points in S.

Definition 4.2.2 Let X be a Riemann Surface. The function f from X to $\mathbb{C}P^1 = \mathbb{C} \cup \{\infty\}$ is called a *meromorphism*, if $f : X \setminus S \to \mathbb{C}$ is holomorphic for some set of isolated points S. We then call S the *poles* of f and denote the set of all meromorphic functions on X by $\mathcal{M}(X)$.

Lemma 4.2.3 *If X is a connected Riemann surface, then the set of meromorphic functions $\mathcal{M}(X)$ is a field.*

PROOF Let us first note that $\mathcal{M}(X)$ is a ring together with the multiplication of functions, since it is a subring of the functions from X to $\mathbb{C}P^1$. It contains the neutral element of multiplication and the product and sum of meromorphic functions are again meromorphic functions. It remains to be checked that for any non-zero $f \in \mathcal{M}(X)$ it holds that the inverse $1/f$ is also in $\mathcal{M}(X)$. For doing that, we need to show that the zeros of f form a set of isolated points. Denote this set by S_0. Assume that there was a $x \in S_0$ that was no isolated point. Then it is a limit point of S_0. Let (U, ϕ) be a chart around x. Then $f \circ \phi^{-1}$ is a holomorphic function on $\mathbb{C}$ and its set of zeros contains a limit point. Then by the indentity theorem for holomorphic functions ([13], page 122), this composition is identically 0 and so f is 0 at least in some neighborhood of x.
Consider the set of points $y \in X$, where f vanishes in a neighborhood of y. This set is open by definition. It is closed because it contains all of its boundary points by the above argument. Since X is connected by assumption and the set is non-empty, so f vanishes on all of X which is a contradiction. Therefore, S_0 is a set of isolated points which makes $1/f$ a meromorphic function, since the set of holomorphic functions is closed under quotients on non-zero sets. □

Now this lemma gives us the opportunity to view a new arising connection between covering spaces and field extensions, because we have the covering spaces of X and the field extensions of $\mathcal{M}(X)$. By the previously developed theories, we will try to bring together these two notions.

Definition 4.2.4 Let X be a connected, compact Riemann surface. Then the non-constant holomorphic map $p : Y \to X$ with finite fibers is a *branched covering*, if $p : Y \setminus p^{-1}(S) \to X \setminus S$ is a finite covering map for a set of isolated points $S \subset X$. Since X is compact, this means that S is finite.

Example: Let us view the holomorphic map $f : \mathbb{C} \to \mathbb{C}$ with $f(z) = z^2$. One can easily see that this map is surjective and the preimage of every point in $\mathbb{C}^\star$ has cardinality 2. However, the preimage of 0 only contains one element, namely 0. Hence, f is not a covering map. Nevertheless, f is a branched covering, because it restricts to a covering map $f : \mathbb{C}^\star \to \mathbb{C}^\star$.

Definition 4.2.5 We define the *degree* of a finite covering to be the cardinality of its fibers. Similarly, we can define the *degree of a branched covering* as the degree of the induced covering map.

Proposition 4.2.6 *Let $L/\mathcal{M}(X)$ be a finite algebraic field extension for a connected compact Riemann surface X. Then there exists a compact Riemann Surface X_L such that $\mathcal{M}(X_L) \cong L$ as an $\mathcal{M}(X)$-algebra.*

PROOF (SKETCH) Further details are in Szamuely, [1], pages 75f.

Using the primitive element theorem 1.3.8, let γ be that primitive element for L and let m_γ be the corresponding minimal polynomial in $\mathcal{M}(X)[t]$. By lemma 1.3.5, m_γ is irreducible and therefore, since m_α is separable, by the euclidian algorithm we can find polynomials A and B in $\mathcal{M}(X)$ such that $A \cdot m_\gamma + B \cdot m'_\gamma = 1$. Here m'_γ denotes the derivative of m_γ. We now evaluate the coefficients of m_γ at $x \in X$, where all of them are holomorphic. We let $S \subset X$ be a set of isolated points of the poles of the coefficient functions of $m_\gamma(x)$ together with the poles of the functions A and B. This implies that on $X' = X \setminus S$, all the functions are holomorphic.
For an open subset $U \subset X'$, let us denote the set of holomorphic functions f on U with $m_\gamma(f) = 0$ by $\mathcal{H}(U)$. Using the theory of a locally-constant sheaf, we get a branched covering X'_L of X. We then need to show that the corresponding Riemann surface X_L is connected and compact and find a function on it satisfying $m_\gamma(f) = 0$. Finally, we prove by mapping f to γ that $\mathcal{M}(X_L) \cong L$.
□

Let $Bra(X)$ be the category whose objects are holomorphic maps from a connected, compact Riemann surface Y to a base Riemann surface X. By $_{\mathcal{M}(X)}SAlg$ we denote the category of finite étale algebras over $\mathcal{M}(X)$. Our aim will be to construct a contravariant functor from $Bra(X)$ to $_{\mathcal{M}(X)}SAlg$.
Assuming we are given a map $\phi : Y \to X$ that is holomorphic, this induces a ring homomorphism $\phi^\star : \mathcal{M}(X) \to \mathcal{M}(Y)$ by $f \mapsto f \circ \phi$. This is a functorial correspondence, since it takes morphisms to morphisms and objects to objects, while preserving the identity morphisms. Let us denote this functor by F. By construction, it is a functor that shall map into the opposite category of the finite algebraic extensions of $\mathcal{M}(X)$.
We can then identify $\mathcal{M}(X)$ with $\phi^\star(\mathcal{M}(X))$ which makes it a finite field extension. This will be proved in the following. We now show that the degrees correspond.

Lemma 4.2.7 *Let $\phi : Y \to X$ be a non-constant holomorphic map of connected Riemann surfaces which has degree d as a branched cover. Every meromorphic function $f \in \mathcal{M}(Y)$ satisfies a polynomial equation of degree d over $\mathcal{M}(X)$.*

PROOF For the proof see Szamuely, [1], pages 73f. □

Proposition 4.2.8 *Let $\phi : Y \to X$ be a non-constant holomorphic map of compact connected Riemann Surfaces that has degree d as a branched covering. The induced algebraic field extension $\mathcal{M}(Y)/\phi^\star(\mathcal{M}(X))$ is finite and has degree d.*

PROOF Let us choose an arbitrary $x \in X \setminus S$. By definition 4.2.5 we know that it has d distinct preimages $y_1, \ldots, y_d \in Y \setminus \phi^{-1}(S)$. Using Riemann's Existance Theorem 4.1.5 we know that there exists a function $f \in \mathcal{M}(Y)$ such that f is holomorphic on all the y_i and has distinct values in $\mathbb{C}$. Using lemma 4.2.7, f satisfies a polynomial equation of degree d, but we can make it an irreducible polynomial equation of degree $n \leq d$ over $\mathcal{M}(X)$

$$\phi^\star(a_n)f^n + \cdots + \phi^\star(a_1)f + \phi^\star(a_0) = 0 \qquad \text{with } a_i \in \mathcal{M}(X).$$

Assuming the a_i are holomorphic at x, we get that the polynomial $a_n(x)t^n + \cdots + a_0(x) \in \mathbb{C}[t]$ has d distinct complex roots, namely the $f(y_i)$, and therefore, using the basic property of polynomials that a polynomial g has at most $\deg(g)$ roots together with our assumption $n \leq d$, we necessarily get that $n = d$. This yields that f is an irreducible polynomial of degree d.

Now, if a_i had a pole at x, it would not be holomorphic. We can easily fix that: There exists a neighborhood of x that does not contain any points of S, as S is a set of isolated points by definition. Since f is holomorphic and has distinct values at all the elements of $\phi^{-1}(x)$, we can by continuity choose another point x', where all the a_i are holomorphic and use the same argument as before. This is possible, because S is finite, since X is compact.

Furthermore, picking any $g \in \mathcal{M}(Y)$ we get for f as above by the theorem of the primitive element 1.3.8 that there exists an $h \in \mathcal{M}(Y)$ such that $\mathcal{M}(X)(f, g) = \mathcal{M}(X)(h)$. Especially, this implies that $\mathcal{M}(X)(f) \subset \mathcal{M}(X)(h) \subset \mathcal{M}(Y)$. However, by lemma 4.2.7 h should also satisfy an irreducible polynomial equation over $\mathcal{M}(X)$ of degree smaller or equal to d. Therefore, the field-inclusion from above is an equality. This means that $g \in \mathcal{M}(X)(f)$ and hence $\mathcal{M}(X)(f) = \mathcal{M}(Y)$. This implies that the field extension from the proposition has degree $\deg(f) = d$. □

Theorem 4.2.9 *Let X be a compact, connected Riemann Surface and $\mathcal{M}(X)$ its field of meromorphic functions. There exists an anti-equivalence of categories between finite field extensions of $\mathcal{M}(X)$ and the branched coverings of X.*

Proof First, the functor F is well-defined by proposition 4.2.8, since it shows that F indeed maps to the category of *finite algebraic* field extensions. By proposition 4.2.6 we know that F is essentially surjective. We still need to show that F is fully faithful. For a connected Riemann surface Y with a holomorphic map $\phi : Y \to X$ that by 4.2.4 is a branched covering of X. This proof is given for finite étale algebras in Szamuely, [1], page 76, so we merely need to restrict it, applying our definition of a branched covering.

Since F is fully faithful and essentially surjective, the categories from the theorem are anti-equivalent by lemma 1.1.9. □

This leads us to the theorem that connects all of the chapters that were treated before on Riemann surfaces.

Theorem 4.2.10 *Let X be a connected compact Riemann surface and let $X' = X \setminus S$ for some set of isolated points S. Let Y_i be connected compact Riemann surfaces such that there is a holomorphic map $\phi_i : Y_i \to X$ that restricts to a covering of X', and such that $\mathcal{M}(Y_i)$ is a finite subextension of $\mathcal{M}(X)^{sep}/\mathcal{M}(X)$. We define $K_{X'} = \bigcup_{i \in I} im(\iota_i)$, where $\iota_i : \mathcal{M}(Y_i) \to \mathcal{M}(X)^{sep}$ is the inclusion map. Then the field extension $K_{X'}/\mathcal{M}(X)$ is Galois and we have for any $x \in X'$:*

$$Gal(K_{X'}/\mathcal{M}(X)) \cong \hat{\pi}_1(X', x)$$

where $\hat{\pi}_1(X', x)$ is the profinite completion of $\pi_1(X', x)$.

Proof For the proof see Szamuely, [1], pages 78f.

Example: Let us view the only example of a compact connected Riemann surfaces that we know at this point and apply the above theorem to get confidence in what it means. This example is the Riemann sphere $\mathbb{C}P^1 = \mathbb{C} \cup \{\infty\}$.

By basic facts of complex analysis, we know that the automorphisms of $\mathbb{C}P^1$ are the Möbius tranformations $\mathbb{C}(t)$ and there exists a unique Möbius transformation $f \in \mathbb{C}(t)$ that maps some distinct fixed points $x_1, x_2, x_3 \in \mathbb{C}P^1$ to some other distinct fixed points $z_1, z_2, z_3 \in \mathbb{C}P^1$, such that $f(x_i) = z_i$. This enables us to pick a finite set $S \subset \mathbb{C}P^1$ of 1, 2 or 3 points in $\mathbb{C}P^1$ and get that $\mathbb{C}P^1 \setminus S$ is biholomorphic to $\mathbb{C}P^1 \setminus S'$ for any $S' \subset \mathbb{C}P^1$ of the same cardinality.

In the following, we will check for $|S| \in \{1, 2, 3\}$, what theorem 4.2.10 says for any of them.

$|S| = 1$: Let us choose the "critical point" ∞ as finite set S. Then $\mathbb{C}P^1 \setminus \{\infty\} = \mathbb{C}$. Since $\mathbb{C}$ is simply-connected and path-connected, it has trivial fundamental group at every point. As the fundamental group then is finite, its profinite completion is also trivial. The theorem shows that $Gal(K_{\mathbb{C}}/\mathbb{C}(t)) = \{0\}$, implying $K_{\mathbb{C}} = \mathbb{C}(t)$. Since $\mathbb{C}$ is simply-connected, if a holomorphic non-constant map $Y \to \mathbb{C}P^1$ is branched only at ∞, then it is an isomorphism.

$|S| = 2$: In this case, if we choose the set $S = \{0, \infty\}$, then we get $\mathbb{C} \setminus S = \mathbb{C}^\star$. With the topological quotient map $\psi : \mathbb{C}^\star \to S^1 = \{z \in \mathbb{C} : |z| = 1\}$, $z \mapsto z/|z|$ we know that $\mathbb{C}^\star$ and S^1 have the same fundamental group. Since $\mathbb{C}^\star$ and S^1 are path-connected, $\pi_1(\mathbb{C}^\star) = \mathbb{Z}$. By a previous example its profinite completion is $\hat{\mathbb{Z}}$. Figure 4.3 portraits the universal covering space $\mathbb{R}$ of $\mathbb{S}^1$, displayed in a suiting way.

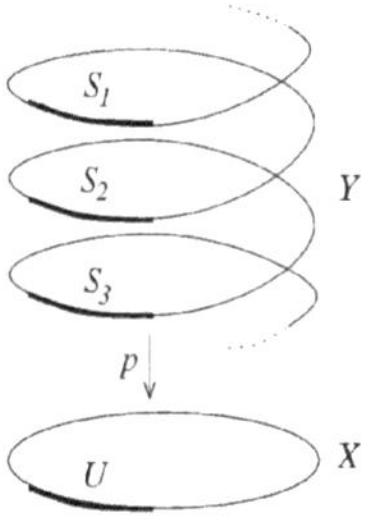

Figure 4.3: The universal covering space of S^1. [3]

Moreover, each finite covering space of S^1 looks similar to this: after a finite amount of rotations, the ends are connected. Since S^1 is a quotient of $\mathbb{C}^\star$, this should give us an idea of how the covering spaces of $\mathbb{C}^\star$ look like, assuming we know that the universal covering space of $\mathbb{C}^\star$ is $\mathbb{C}$. By theorem 4.2.10, this also means that $Gal(K_{\mathbb{C}^\star}/\mathbb{C}(t)) = \hat{\mathbb{Z}}$ and that the Galois group of any subextension of $\mathbb{C}(t)$ is isomorphic to $\mathbb{Z}/n\mathbb{Z}$ for some $\mathbb{Z}$.

This implies that all coverings that are branched at 0 and ∞ are of the form $p : \mathbb{C}P^1 \to \mathbb{C}P^1$ with $t \mapsto t^n$, since then $Aut(p) = \mathbb{Z}/n\mathbb{Z}$. Here, the automorphisms are $t \mapsto \xi^j t$, where $\xi = e^{2\pi i/n}$ and $j \in [n]$.

$|S| = 3$: This time, let us choose $S = \{0, 1, \infty\}$. The fundamental group of $\mathbb{C} \setminus \{1, \infty\}$ is the free group with two generators $\mathbb{Z} \star \mathbb{Z}$. Obviously, the universal cover of $\mathbb{C} \setminus \{1, \infty\}$ is $\mathbb{C}$ and the profinite completion of the fundamental group is $\widehat{\mathbb{Z} \star \mathbb{Z}}$, so by theorem 4.2.10 the Galois group of the collection of all field extensions of meromorphic functions that have finite degree over $\mathbb{C}(t)$ and correspond to coverings of $\mathbb{C} \setminus \{1, \infty\}$ is isomorphic to $\widehat{\mathbb{Z} \star \mathbb{Z}}$.

[3] https://de.wikipedia.org/wiki/Rotationszahl

Bibliography

[1] Tamás Szamuely, *Galois Groups and Fundamental Groups.* Cambridge University Press, 2009.

[2] James R. Munkres, *Topology.* Pearson, 2000. Second Edition.

[3] Melissa Lynn, *Galois Categories.*
`https://www.math.uchicago.edu/~may/VIGRE/VIGRE2009/REUPapers/Lynn.pdf`, 2009.

[4] James S. Milne, *Fields and Galois Theory.*
`http://jmilne.org/math/CourseNotes/FT.pdf`, 2017. Version 4.53.

[5] Hendrik W. Lenstra, *Galois Theory for schemes.*
`http://websites.math.leidenuniv.nl/algebra/GSchemes.pdf`, 2008. Third Edition.

[6] Askold Khovanskii. *Galois Theory, Coverings and Riemann Surfaces.* Springer-Verlag, 2013.

[7] J. Peter May, *A Concise Course in Algebraic Topology.* CreateSpace Independent Publishing Platform, 2013.

[8] Richard Koch, *Existence of a Universal Cover.*
`http://pages.uoregon.edu/koch/math432/Universal_Cover.pdf`, 2006.

[9] David Glickenstein, *Covering Spaces.*
`http://math.arizona.edu/~glickenstein/math534_1011/coveringspaces.pdf`, 2011.

[10] Klaus Lamotke, *Riemannsche Flächen.* Springer Verlag, 2009. Zweite Auflage.

[11] The Stacks Project, *Galois Categories.*
`https://stacks.math.columbia.edu/tag/0BMQ`, 2015.

[12] Alexandre Puttick, *Galois Groups and the Étale Fundamental Group.*
`https://webusers.imj-prg.fr/~jean-francois.dat/enseignement/memoires/M1AlexPuttick.pdf`, 2012.

[13] Mark J. Ablowitz; Athanassios S. Fokas, *Complex variables: Introduction and Applications.* Cambridge University Press, 1997.